POWER METAL

POWER
METAL

The Race for the Resources
That Will Shape the Future

VINCE BEISER

RIVERHEAD BOOKS   NEW YORK   2024

RIVERHEAD BOOKS
An imprint of Penguin Random House LLC
penguinrandomhouse.com

Copyright © 2024 by Vince Beiser
Penguin Random House values and supports copyright. Copyright fuels
creativity, encourages diverse voices, promotes free speech, and creates
a vibrant culture. Thank you for buying an authorized edition of this
book and for complying with copyright laws by not reproducing, scanning,
or distributing any part of it in any form without permission. You are
supporting writers and allowing Penguin Random House to continue
to publish books for every reader. Please note that no part of this book
may be used or reproduced in any manner for the purpose of training
artificial intelligence technologies or systems.

Riverhead and the R colophon are registered trademarks of
Penguin Random House LLC.

LIBRARY OF CONGRESS CATALOGING-IN-PUBLICATION DATA

Names: Beiser, Vince, 1965– author.
Title: Power metal : the race for the resources
    that will shape the future / Vince Beiser.
Description: New York : Riverhead Books, 2024. |
    Includes bibliographical references and index.
Identifiers: LCCN 2024011709 (print) | LCCN 2024011710 (ebook) |
    ISBN 9780593541708 (hardcover) | ISBN 9780593541722 (ebook)
Subjects: LCSH: Mines and mineral resources—Environmental aspects. |
    Mineral industries—Social aspects. | Strategic materials—
    Economic aspects. | Sustainable development—Forecasting. |
    Rare earth metals—Government policy.
Classification: LCC TD195.M5 B44 2024 (print) |
    LCC TD195.M5 (ebook) | DDC 333.8/5—dc23/eng/20240429
LC record available at https://lccn.loc.gov/2024011709
LC ebook record available at https://lccn.loc.gov/2024011710

Printed in the United States of America
1st Printing

Book design by Daniel Lagin

For Edith and Elaine
And Jed and Jane.

Greatly admired
Deeply missed.

CONTENTS

PART THREE: Better Than Recycling

There's No Such Thing as Clean Energy

bought my first all-electric car in 2018, back in what felt like the pioneer days of renewable energy, and, man, did I feel virtuous. There I was, with my unglamorous but morally exemplary used Nissan Leaf, doing my part to save the Earth.

At the time, I lived with my family in Los Angeles. Ours was a fairly typical setup—two parents, two kids, cat, dog, and, of course, two cars. How else could we have gotten around in a city like LA? But I was also very concerned about climate change, which was already afflicting life in our area with increasingly frequent and potent water shortages, wildfires, and heat waves. So, when our old Mazda hatchback finally gave out, I made the leap from gas to electric. At last, I thought, I was on the right side of history! No longer was I supporting any planet-plundering oil companies. My car emitted not a single ounce of carbon, no matter how far I drove. It ran on a battery I charged from a regular electrical outlet, just like my laptop and iPhone. In fact, I was pleased to learn, my car's battery, made mainly of lithium, cobalt, and nickel, was basically just a bigger version of the batteries in my digital devices.

True, the electricity to power those batteries came mostly from fossil-fueled power plants. But I expected that those carbon-spewing behemoths would soon be replaced by renewable sources, like solar and wind farms. Once that energy transition was accomplished, I figured, I would be leading an environmentally blameless, climate-friendly life. My car would run silently on invisible electricity. I'd work and communicate with people wirelessly. The energy for it all would be drawn from sunlight and thin air. The image I had of my digitally enabled, renewably powered future life had a wonderfully weightless, dematerialized, untethered feeling of freedom to it.

Unfortunately, it was also utterly wrong.

As a journalist, it's my job to be inquisitive, to ask questions. I was curious about where the lithium, cobalt, nickel, and other raw materials for all those batteries in my machines came from. I started doing some research and quickly came to realize that the electric car I was so proud of, the digital devices I use every day and the renewable energy I was counting on to power them, are together spawning massive environmental damage, political upheaval, mayhem, and murder. To get the raw materials required to build cell phones, electric cars, and wind turbines, rainforests are being cut to the ground. Rivers are being poisoned. Children are being put to work in mines. Warlords and a billionaire crony of Vladimir Putin are getting rich. And untold numbers of people are getting killed.

The human race is facing a paradox: We must do everything we can to stave off the catastrophes of climate change, but, in doing so, we may create a whole other set of catastrophes. This book is an attempt to make clear the extent of the damage that process is already inflicting on people and the planet, the many ways in which it might get worse—and how we can do better. To build a sustainable world run on the digital technology we already take for granted and the carbon-free power we absolutely require, we need a whole new approach.

POWER METAL

CHAPTER I

The Electro-Digital Age

W e are moving into a new era, a new phase in humanity's economic and social development. The twenty-first century is being shaped by forces that are changing how we communicate with each other, how we travel from place to place, how we heat and cool our homes, how we live our lives on an increasingly stressed planet. There are three major, interconnected drivers of this era: digital technology and the internet; renewable energy; and electric vehicles. Call it the Electro-Digital Age.

Digital tech, of course, is already deeply embedded in our lives and only growing more essential to almost everything we do. The other two drivers—renewable energy and electric vehicles—are coming on fast. In many ways, all of that is good news. The energy transition from fossil fuels to renewables is a crucial part of the cure for climate change. But it's a cure with brutal side effects. Taken together, the three pillars of the Electro-Digital Age are inflicting tremendous but largely overlooked harm all around the planet, including environmental catastrophes, child labor, slavery, robbery, and murder. They are also altering the balance of geopolitical power, mostly to the benefit of authoritarian regimes not especially friendly to the West. All of which is largely because the three pillars of this new, high-tech era depend on a set of ancient, natural materials: metals.

In our deeply digital, relentlessly online modern world, it's easy to forget, or ignore, or fail to appreciate the extent to which our lives depend on the extraction, transportation, and processing of titanic quantities of physical materials. Millions of us, myself included, make our living dealing only with dematerialized abstractions, intellectual "products" that exist mainly on computers—articles, analyses, marketing plans, software, videos, podcasts. We process financial data, we build websites, we run social media campaigns, we develop business plans—products that have no physicality, no corporeality to them, beyond perhaps a hard copy version printed on paper. Information streams into our offices, we process it into other forms of information, and it streams back out again.

But all of that is made possible by machines—computers, phones, wireless routers, internet cables, data centers packed with computer servers—all of them drawing energy from enormous power plants. Just as most of us are disconnected from the processes by which our food is produced, we are disconnected from the processes by which our machines are manufactured. The raw materials from which those machines are built make our lives possible. And they come at enormous expense.

Laptops, tablets, and cell phones are made from a kaleidoscopic array of materials, from everyday metals to far more exotic and obscure substances. Mobile phones can contain as many as two thirds of all the elements in the periodic table, including dozens of different metals. Some of those metals are familiar: There's gold in the typical phone's circuitry, tin in its circuit board, nickel in its microphone. Some are not. Tiny flecks of indium in the screen make it sensitive to the touch of your finger. Europium enhances the colors you see on that screen. Neodymium, dysprosium, and terbium are used to build the tiny mechanism that makes your phone vibrate.

The batteries that power cell phones—probably the components to which we give the most thought—are made with lithium, cobalt, and nickel. Similar batteries energize your rechargeable drill, Roomba, electric toothbrush, and countless other cordless electric devices. That includes most electric vehicles, from e-scooters to Tesla SUVs. Electric cars are battery cars, but their batteries are *big*. Just one Tesla Model S can contain as much lithium as ten thousand mobile phones.

All of those batteries need to be charged, and charged again and again, with electricity. The more batteries there are, the more electric power we must generate to feed them. The millions of new electric vehicles hitting the world's roads each year come with a monstrous appetite for power. Driving those electric cars doesn't create carbon emissions, but generating the electricity that powers them often does. Today, the world's biggest source of electrical energy is still coal-fired power plants. The catastrophic impacts of our reliance on such carbon-intensive fuels are well-known. To keep those impacts to a minimum, we need to switch not only to electric cars but also to carbon-neutral renewable-energy sources, especially solar and wind, to power them.

Sunlight and air, beloved elements of the natural world, seem like wonderfully benign energy sources. So much more appealing than filthy coal or gooey petroleum. Solar and wind power are often described as free energy, since they literally fall from the sky. But they entail a Faustian bargain. To capture energy from the sun and wind, transmit it, and use it, we need machines. We need wind turbines, solar panels, switching stations, power lines, batteries.

Picture what's involved in the process of generating, transmitting, and using renewable energy. Los Angeles, for instance, gets a little more than 4 percent of all its electricity from the Red Cloud wind farm southeast of Albuquerque, New Mexico. The farm's dozens of towering turbines are made out of steel reinforced with a rare metal called niobium. Wind hits a turbine's blades, making them rotate. Inside the turbine, special magnets made with neodymium, another metal, convert that movement into electric power. Cables made with thousands of pounds of aluminum and copper then carry that electricity from the turbine into the New Mexico power grid. The electricity flows along hundreds more miles of copper cables to a switching station in Arizona and on from there to thousands of homes, garages, and offices in Los Angeles.

Thanks to those cables delivering the juice to my house, I can plug in my Leaf and electricity will flow into its battery. That battery is made with approximately one hundred pounds of nickel, cobalt, and lithium, and nearly as much copper. When I step on the accelerator, hundreds of yards of coiled wires, made out of more copper, activate neodymium-based magnets, similar to the ones in the wind turbines, to convert the electricity back into movement, turning the car's wheels.

All of those machines, cables, wires, and batteries are made with metals. Metals don't fall from the sky. They are ripped from the Earth.

The Electro-Digital Age demands a terrifying amount of such ripping. To manufacture all the digital tech we demand and all the electric cars, wind turbines, cables, magnets, and other gear we require for the transition to renewable energy, we're going to need titanic quantities of what are variously called battery metals, technology metals, transition metals, or the term I prefer, "critical metals." In all of human history, we've mined about seven hundred million tons of copper. We'll need to mine the same amount again in the next twenty-odd years. By 2050, the International Energy Agency estimates, demand for cobalt from electric vehicle makers alone will surge to nearly five times what it was in 2022; nickel demand will be ten times higher; and for lithium, fifteen times higher, the annual total soaring from just under seventy thousand tons to over one million. The surge is well underway. The market size for copper, cobalt, and rare earths nearly doubled between 2017 and 2022, tripled for nickel, and septupled for lithium, reaching a total of $320 billion.

"Energy transition minerals, which used to be a small segment of the market, are now moving to center stage in the mining and metals industry," declared a 2023 IEA report. The market for critical metals is set to continue rising in the coming years—assuming the world can produce them. "The prospect of a rapid increase in demand for critical minerals—well above anything seen previously in most cases—raises huge questions about the availability and reliability of supply," warns the IEA.

All over the world today, governments, corporations, entrepreneurs, activists, and researchers are racing to figure out how to meet that mushrooming demand. Those millions of tons of metals have to come from somewhere. Today, one industry provides the overwhelming majority of them. Our high-tech, carbon-free future depends on one of humanity's oldest and dirtiest endeavors: mining.

Mining is a rough business. Metals are natural resources, products of the Earth, but the Earth doesn't give them up willingly or easily. Digging up metals typically involves destroying the Earth in the most literal sense. The whole object is to tear up trees or grasslands or deserts, blast apart the underlying rock and earth with explosives, and rip out the remains. And it doesn't stop there. The metal-bearing ore clawed out of the ground has to be processed, smelted, and refined with enormous, energy-guzzling, pollution-spewing industrial equipment and oceans of chemicals.

"Mining done wrong can leave centuries of harm," says Aimee Boulanger, a former antimining activist who now heads the Initiative for Responsible Mining Assurance, a group that works with mining companies to develop sustainable practices.

The range and extent of damage inflicted by mining is staggering. Metal mining is America's leading toxic polluter and has sullied the watersheds of almost half of all the rivers in the American West. Chemical leaks and runoff often foul the air and water around metal mines. Mines also generate enormous piles of poisonous waste, which are stored behind dams that have a terrifying tendency to fail. Torrents of toxic sludge pouring through collapsed tailings dams have poisoned rivers and lakes from Canada to Brazil and killed hundreds of people. Those casualties are on top of the hundreds, possibly thousands, of miners who die in workplace accidents each year.

Chemical leaks and dam failures are unintended consequences. But even when everything is going as it's supposed to, mines inevitably inflict damage. They devour resources and excrete waste on a titanic scale. A study from the Vienna University of Economics and Business found that industrial mines have wiped out more than one thousand square miles of forests since 2000. Valuable metals typically make up only a tiny fraction of all the rock and earth that has to be dug up to get them. To get just one ton of nickel, an average of two hundred fifty tons of ore and waste rock must be processed. For copper, it's twice that much. Seventy-five pounds of ore have to be wrested out of the earth to build a single four-and-a-half-ounce iPhone. That ore has to be crushed and the metals separated from the waste rock around them, industrial processes that belch out as much as one hundred pounds of carbon per phone.

Mines also suck up huge quantities of water, a major point of conflict in arid regions such as northern Chile, which is home to some of the world's most immense copper and lithium operations. And the mines require fleets of drill rigs, trucks, diggers, and other heavy machinery, energy hogs that erupt megatons of greenhouse gases—as much as 7 percent of the world's annual total.

None of this is exactly welcomed by whoever lives nearby. Irate local residents and Indigenous communities are fighting proposed critical-metal mines all across the United States, from North Carolina to Nevada, as are their counterparts in Canada, Serbia, Brazil, the Philippines, and many other countries. In some places, this kind of opposition can have lethal consequences. At least 320 antimining activists around the world have been murdered since 2012. And those are just the ones Western human rights groups know about.

To keep our technology-abetted lifestyles going and to kick our fossil-fuel dependence, we humans need to get our hands on more critical metals. But we also need to find cleaner, more humane, more sustainable ways of getting those metals. How can we do that? Can we do it soon enough? Can we make our way to a truly sustainable world without trashing the planet in the process?

Searching for answers to those questions took me to mines, protest sites, and research labs around the world. I visited a shipyard in Belgium, a desert in Chile, a junkyard in Canada, and the biggest garbage dump in West Africa. I met some of the wild array of wealthy investors, grassroots activists, scientists, politicians, manual laborers, and artificial-intelligence experts who are all players in the trillion-dollar global race for the metals we need for this new era.

It's a race unfolding across the planet—and even on the ocean floor, not to mention outer space. Some aspects of it are genuinely promising, some are smoke-and-mirrors schemes, and some are disturbingly dangerous. This race will not only reshape industries, it will also affect the destinies of entire countries and alter the global balance of power. And it will force us to question how we organize our cities, our societies, and our lives.

The digital revolution is by now part of the everyday lives of almost everyone in the Western world, and, increasingly, the lives of everyone on Earth. As of January 2024, an estimated 5.35 billion people—about two thirds of the world's population—were using the internet. Nearly as many own mobile phones. All told, there are more than fifteen billion mobile devices in use worldwide.

The shift toward renewable energy and electric vehicles is also happening much faster than most people realize. Humanity is on the cusp of a new energy era that could be as transformative as our shift from wood to coal, and from coal to oil. Today's children will inherit a world powered in ways fundamentally different from that of their parents or grandparents. The shape of power, in the literal sense, is changing profoundly.

In talking about renewables, I'm focusing mainly on wind and solar power, because, at this point, they are the most advanced and rapidly expanding new sources of renewable electricity. There are others, of course. Hydropower is clean and renewable, but we've already dammed about as many rivers as we can (or should). Worse, climate-change-induced droughts are drying up rivers and reservoirs around the world, imperiling the hydro-energy supply. Many people consider nuclear power a viable, carbon-neutral alternative to fossil fuels, but that's a deeply complicated issue beyond the scope of this book (plus, it's not really renewable). Hydrogen may become a major energy source someday, but that day is still a long way off.

Solar and wind, meanwhile, are exploding. At the start of the twenty-first century, wind and solar combined produced a fraction of a percent of all the world's electricity. Now, they provide more than 12 percent. In the United States, renewables now provide almost a quarter of the nation's electricity, outstripping both coal and nuclear. The IEA expects America's solar and wind capacity to nearly double by 2027. As of 2022, wind and solar were generating more power than all the world's nuclear plants. The disruptions to Russian oil and gas exports brought on by the invasion of Ukraine in February of 2022 jolted many nations into redoubling their efforts to develop more dependable alternatives, sparking what the IEA calls "unprecedented momentum for renewables." In 2023, the world created enough new renewable energy to power all of Germany and Spain. By 2027, the IEA predicts renewables will be the single largest source of electricity worldwide.

The electric car market is growing even faster—so fast that it's reshaping the entire automobile industry. As recently as 2012, only 120,000 electric vehicles were sold worldwide each year. By 2022, customers were snapping up more than that number every *week*. Sales of new electric vehicles are expected to top 30 million by 2030. California, Washington, and Oregon have all declared they will ban the sale of new internal combustion vehicles by 2035, and at least twenty countries have announced similar future prohibitions.

Every major automaker is rolling out electric models, from Cadillacs to pickup trucks. Industry heavyweights, including General Motors and Volkswagen, have declared they aim to phase out petroleum-powered vehicles altogether in the coming years. Electric-car battery factories are opening all over the United States and around the world. To safeguard their supply chains, carmakers are deepening their involvement with the mining industry. GM, for instance, recently sank $650 million into a proposed lithium mine in Nevada.

The critical-metal mining boom is also shaking up geopolitics. That's natural. Think about the last major energy transition, when the world shifted from coal to oil. Back in the coal-powered days, Saudi Arabia was an afterthought in global affairs. But with the rise of the automobile, the world suddenly needed huge amounts of a substance that hadn't been in much demand before: oil. Almost overnight, Saudi Arabia became one of the richest nations on the planet. (Having learned from their history, the desert kingdom is now also investing heavily in critical-metal production.)

Similarly, a handful of little-noticed nations from the Arctic to the South Pacific now have a chance to reap fortunes from their huge reserves of metals that practically nobody cared about until recently. Way out in the South Pacific, the tiny French territory of New Caledonia holds as much as a quarter of all the world's unmined nickel. In the high Arctic, Greenland has enormous quantities of rare earths. Bolivia, in the middle of South America, has the world's largest deposits of lithium. The Democratic Republic of the Congo, in the heart of Africa, holds almost half of the world's cobalt reserves. Afghanistan harbors enormous deposits of copper, cobalt, and other metals. Remember the niobium that reinforced those wind turbines? Brazil produces almost all of the world's supply.

In Southeast Asia, Indonesia and the Philippines are just starting to fully exploit their huge nickel reserves. For now, however, the world's top producer of high-grade nickel is Russia. It's also a huge exporter of copper and other metals. Fears that Russia's war in the Ukraine might affect its nickel production sent the metal's prices skyrocketing in the days after Vladimir Putin launched the invasion in February 2022. Turns out, traders need not have worried. Though most Russian exports were closed off by international sanctions, the world quietly decided that Moscow's nickel, copper, palladium, and other minerals were too important to shut off. For more than two years after the war began, Russia continued to export billions of dollars' worth of nickel to the West. In other words, to some extent, the switch to electric cars helped to fund Russia's invasion of Ukraine.

But when it comes to winners in the critical-metals race, one country is way out in front of all the others. No matter what material you're talking about, at least one and quite possibly *all* of the steps in the production chain, from mining to processing to refining to manufacturing the final product, will take place in China.

Leveraging its own natural resources, relatively lax environmental standards, diplomatic clout, and shrewd overseas investments, China has, in recent decades, come to dominate the entire supply chain for critical metals. China has huge reserves of lithium and other metals, some of which it allegedly mines with forced labor. The homegrown resources it lacks, it buys abroad; Chinese companies own mines all around the world that produce raw cobalt, nickel, and many other metals.

Regardless of where critical metals are dug up, or by whom, most will end up sent to China for refining and processing. China has more than half the world's refining capacity for lithium, cobalt, and graphite (another key battery ingredient) and close to that much for nickel and copper. Other Chinese factories then take those refined metals and turn them into most of the world's solar panels, a hefty share of its wind turbines, nearly three quarters of all lithium-ion batteries, and a majority of all electric vehicles. All of which gives Beijing not only a commanding position in the emerging economy of the Electro-Digital Age but also enormous geopolitical leverage. China has already shown, in recent years, that it is willing to cut off world supplies of a particular set of critical metals to support its political goals. It could unsheathe the embargo weapon again at any time.

Western nations have belatedly awakened to this vulnerability and are scrambling to address it. The United States, Canada, Japan, and the European Union have all explicitly prioritized finding non-Chinese sources for critical metals and are pouring cash and resources into the quest. "The United States' mineral import dependency and the concentration of mineral supply from certain countries are broadly recognized as growing threats to economic growth, competitiveness, and national security," warned the US Senate's Committee on Energy and Natural Resources in 2019. In 2022, the US Congress enacted an infrastructure package that included $7 billion to expand the domestic supply chain for battery minerals. That same year, it passed the Inflation Reduction Act, which includes many more billions of dollars to subsidize batteries and electric vehicles made with domestically sourced metals.

The good news is that there are many ways the rest of the world can get its hands on the metals it needs without bolstering Beijing (or Moscow) or devastating natural landscapes. Start in the most obvious place: the mining industry itself.

I knew practically nothing about mining when I started researching this book, but I assumed it was a dirty, destructive industry, disastrous for the environment and anyone living nearby. Historically, that has often been true. But, to my surprise, I learned that it's not as true now as it used to be. The industry and the context it operates in have changed a lot in recent years and continue to change, often for the better.

In decades past, mining companies in league with greedy governments could just pick a spot to dig, shove aside whoever and whatever was living there, gouge the metals out of the ground, and dump the waste wherever was handy. Today, that kind of rape-and-run approach is much more difficult to get away with. Stricter government regulations, higher environmental and social standards, and the industry's evolving perception of what's in their self-interest are doing a lot to change mining practices and minimize their damage. Much of that is because mining's chief opponents—environmentalists and the local communities and

Indigenous people who bear the brunt of the industry's impacts—now have far more legal protection, political power, and social clout than they used to.

Fifty years ago, the environmental movement barely existed. Indigenous peoples and local communities had little recourse against mining companies that came looking for precious metals on their lands. The playing field was tipped wildly against them. Today, that field is still far from level, but it's not nearly as lopsided as it used to be. Groups like Greenpeace, Friends of the Earth, and the World Wildlife Fund have millions of members and operations in dozens of countries. A welter of international agreements and national regulations are on their side. And in a world in which almost everyone has a video camera in their pocket and the means to broadcast to the world, it's much harder to get away with the terrible practices for which the industry had become infamous. Even industry titans like Robert Friedland, a Canadian American mogul nicknamed "Toxic Bob," acknowledge this. "Every one of these hand phones is an NGO," he told a major copper-industry conference in Chile in 2022—meaning that every cell phone is a potential non-governmental organization, one of the activist groups that so complicate life for industries like his. "Click! And you're on the cover of *The New York Times*. You can't hide anymore."

Indigenous people across the globe have much more political representation and legal protections than they used to. In April of 2022, I attended the First Nations Major Projects Coalition's annual conference, a gathering of Canadian Indigenous leaders and the companies doing business on their lands. It was a high-end corporate event held in a swanky Vancouver hotel, attended by executives from some of the world's top mining companies. The fact that such an event even exists speaks to how much more power Indigenous people now hold and how skillful they have become at leveraging it. Indigenous people didn't even have the right to vote in Canada until 1960.

Consumers are also applying much more pressure than ever before. The growing consciousness about the origins of the products that people buy and use every day—from fair-trade coffee to dolphin-safe tuna to sustainable clothing—is spreading to mining. It's late in coming, partly because most people rarely think about the industry or its role in manufacturing our cell phones and solar panels. Hardly any of us could name even the most important companies involved, such as Anglo American or Glencore. Why would we? We're not their customers; other businesses are. But those businesses include household names like Tesla and Apple. Social-justice and environmental groups are ratcheting up the pressure on those companies to make sure their critical-metal supply chains don't run through slave camps or clear-cut forests.

All of that adds up to actual clout. In 2020, for instance, Rio Tinto, one of the world's biggest mining companies, blew up a historic Aboriginal site in Australia to build an iron mine. A century ago, that would have been business as usual. In 2020, that destruction sparked public outrage so fierce that it cost the company's CEO and several other top executives their jobs and forced the company to make restitutions to an Aboriginal group.

"Mining companies know there's a change that's brewing. It started maybe fifteen years ago with attention around blood diamonds or dirty gold," says Boulanger. "Now this stuff that industrialized societies use every day, our phones or buildings or electronics, are really getting this attention. The energy transition has caused that."

All the big miners and manufacturers now speak the language of sustainability, of net-zero emissions and respect for Indigenous peoples and the environment. Even at industry events, where people inside the business are mostly talking to each other, PowerPoint presentations and trade-show banners trumpet the need for sustainability, concern about climate change, and support for empowering communities. "We need to let people know that today's mining industry is not like their father's or grandfather's," a representative of BHP, one of the world's biggest mining companies, told an industry conclave in Vancouver in 2023. "And we have to live that."

Some of that, of course, is just public relations, often called greenwashing, but there are certainly some in the industry who believe it. If nothing else, in today's world, it's in a company's self-interest to limit the damage it causes to the environment and local communities, to preserve what people in the industry call their "social license" to operate, if only to avoid costly lawsuits. It's generally cheaper and easier to build a water-treatment plant or an air-quality monitor at the start of a major new project than to have to go back years later and retrofit all the machinery in response to a court order or public pressure. Let alone to clean up after a major industrial accident. As the historian Jared Diamond put it in *Collapse*, "Cleaning up pollution is usually far more expensive than preventing pollution, just as doctors usually find it far more expensive and less effective to try to cure already sick patients than to prevent diseases in the first place by cheap, simple public health measures."

That growing awareness is real progress. But it also points out another key truth: *Everything* has a cost. It's a cliché, but that doesn't make it any less true. There are no solutions, no technologies, no social or economic developments that bring only benefits. Every development, however positive, also has some kind of downside. There are always winners and losers. The shift to renewable energy will ultimately benefit *most* people, but, in the process, it will impose a steep price on *some* people.

Even the measures that we take to reduce those costs incur other costs. The downside to the upwelling in concern for the environment and human rights is that those concerns slow down the hunt for new sources of critical metals. All those regulations and public hearings and protests and legal challenges mean that it takes many years, often decades, to go from discovering a new deposit to opening up a mine that will bring that metal to market. In the 1950s, it took only three or four years to bring a new copper mine online in the United States; today the average is sixteen years.

"The long lead times for new mining projects pose a serious challenge to scaling up production fast enough to meet growing mineral demand for clean energy technologies," the International Energy Agency warned in a 2022 report. "Current supply and investment plans remain well short of meeting the growing demand from clean energy technologies if the world is to achieve the goals in the Paris Agreement." (That's the 2016 treaty under which almost all the world's nations agreed to try to limit the planet's temperature increase to 1.5 degrees Celsius above pre-industrial levels.) In other words, we might not have enough metals to build wind turbines, solar panels, and electric cars as quickly as we need to in order to keep the world from warming to the point after which climate change could become truly cataclysmic. Such supply shortages could drive up prices, which could, in turn, reduce sales, which "could derail or delay the energy transition itself," warns the International Monetary Fund.

Understand: I am *not* arguing that governments should roll back environmental regulations to make life easier for mining companies. I am pointing out that we are going to have to accept some trade-offs to keep the energy transition moving. Everything has a cost.

In my research for this book, I talked to all kinds of people who are deeply and justifiably concerned about the threats that mining poses to the planet. Those conversations would inevitably come around to some version of me saying, "Yes, agreed, mining causes bad things. This lithium project may damage a desert's ecosystem. That nickel mine may ruin the lives of nearby villagers. But what's the alternative? We need these metals to build the renewable energy systems that will stave off the biggest bad thing of them all, which is climate change. How can we do that without digging more mines?"

Almost always, the answer was, "Recycling!"

When I heard that word, I thought of those plastic bins you put your bottles, cans, and used paper into. You set it all out with your trash, and it's gone the next day. What's so hard about that?

A lot, it turns out. Recycling metal is a completely different proposition from recycling the paper and glass we put in those bins. The business of taking any manufactured product, be it a toaster, a cell phone, or a length of electric cable, and breaking it back down into the raw materials from which it was constructed is a fiendishly complex endeavor. It requires many steps carried out in many different places. Manufacturing those products required an international supply chain: gathering raw materials in one country, processing them in another, shipping them to a factory in a third country where they were turned into new products, and finally selling those products somewhere else altogether. Recycling those manufactured products requires a *reverse* supply chain that is almost as complicated.

Recycling helps. But as a solution, it's utterly inadequate. While recycling does save energy and can reduce the amount of raw materials we use, it too comes with tremendous costs. Some of the processes used to recycle metal, especially in developing countries, are carried out by desperately poor people working in dangerous conditions that generate toxic by-products and lethal pollution.

We tend to think of recycling as the best alternative to using virgin materials. In fact, it's one of the worst. Recycling is the most difficult and energy-intensive way to get further use out of just about any given product. Consider a glass bottle. To recycle it, you have to smash it to pieces, melt down the crushed bits, and mold them into a whole new bottle. It's a whole industrial process that requires a lot of energy, time, and expense.

Or you could just wash out the bottle and reuse it.

Reuse: that's a much better alternative. It's not a new idea. For much of the twentieth century, gas stations, dairies, and other companies that sold things in glass bottles would collect, wash, and fill them up again. The "Reduce, Reuse, Recycle" slogan and its famous three-chasing-arrows logo have been around since the 1970s. Today, they are more relevant than ever.

Rendering an old cell phone, a car battery, or a solar panel down to its constituent metals requires far more energy, labor, and cost than refurbishing that product so that it can be used again. This already happens in rich, developed countries: You can buy refurbished computers, phones, and even solar panels online and in some stores. But where the practice is really widespread is in the developing world. There is a huge market in Africa and Asia for the West's castoffs. If you live in North America, you might not be satisfied with your old iPhone 8 anymore, but there are plenty of people in less affluent countries who would be happy to use it.

There are important lessons for the West there. Perhaps the most important one is this: We must think beyond simply replacing fossil fuels with renewables. We must reshape our relationship to energy and natural resources altogether. That seems like a tall order. And yet, in many ways, and often in unexpected places, that process is already underway. There is a whole range of things we can do—as consumers, as voters, as human beings—to lessen the effects of the electro-digital transition. One, in particular, can make a huge difference.

The critical metals that will enable the electro-digital transition won't come from a single source. They will come from mines of many different types in many different places. They'll come from scrapyards and recycling centers around the world. And some of them will be extracted from entirely new places using completely new methods and technologies. The choices we make about where and how we get those metals, about how much of them we really need, about who will prosper and who will suffer as a result, are tremendously important. Every choice will entail a cost, but some will cost much more than others.

We're lucky in one way: We're just at the beginning of this historic transition. The challenge now is figuring out how to make that transition without repeating the worst mistakes of the last one.

PART ONE

ELEMENTS OF THE FUTURE

|||||||||||||||||||||||||||||||||||||||||||||||||||||||||||||||||||||||||||||||||||||||||||||||||||||||||||||||||||||||||||||||

Everything I do is metal. When I clean my house, it's metal.

—SCOTT IAN, GUITARIST, ANTHRAX

CHAPTER 2

The Elemental Superpower

Shirtless, smoking, and (it was later alleged) drunk at around 9:30 one September morning in 2010, Zhan Qixiong stared across the rolling waves at the Japanese Coast Guard ship bearing down on his fishing trawler. Zhan, a lanky forty-one-year-old wearing denim cutoffs, a white T-shirt, and flip-flops, couldn't have been too surprised. The ship he captained, the *Minjinyu 5179*, was sailing provocatively close to a cluster of tiny, uninhabited but nonetheless fiercely contested islands in the East China Sea. Japan controls them, but China claims them.

Typically, Japanese authorities would simply escort a trespassing ship out of the area. But there was nothing typical about the confrontation that morning. It would set off reverberations that would be felt around the world.

Japan took over the islands back in 1895, during the first Sino-Japanese war, and China has long wanted them back—especially since oil and gas deposits were discovered under the nearby waters in the late 1960s. Tensions have flared several times over the years when Chinese boats have encroached into the islands' waters. In 1978, thirty-eight Chinese fishing ships, some equipped with machine guns, anchored near the islands, their crews shouting and waving signs supporting China's territorial claims. In an effort to head off more serious clashes, the two nations struck an agreement under which no ships from either country are allowed within twelve nautical miles of the islands. But here, more than thirty years later, were Zhan and his crew in his bright blue, thirty-seven-meter trawler tossing lightly on the swell, fishing nets impudently in the water.

The Japanese cutter, named the *Yonakuni*, came up alongside. "You are inside Japanese territorial waters," barked a voice in Chinese over the ship's loudspeakers. "Leave these waters."

Zhan wasn't impressed. Instead of leaving, he turned the trawler sharply, ramming into the *Yonakuni*'s stern at a near-right angle. "It collided with us!" shouted one of the Japanese crew. The *Minjinyu* then sailed off. Soon, a second Japanese boat came racing across the water toward it, sirens wailing, loudspeakers bellowing orders to leave. Zhan watched them come, standing on the deck with a few of his crew, a cigarette in his mouth. When the Japanese got close enough, he again slammed his trawler into their vessel.

Furious Japanese authorities soon boarded Zhan's ship and arrested him and his whole crew. The crew was sent back to China, but Zhan was held and ordered to stand trial. When this news reached China, it sparked a furor. Memories of Japan's brutal invasion during World War II still run deep. Angry crowds protested in the streets of Chinese cities. Chinese premier Wen Jiabao personally called for Zhan's release. The Japanese didn't budge. Then, on September 21, China took drastic action: It cut off exports to Japan of rare earth metals.

Rare earths are among the most important of all the critical metals required for the Electro-Digital Age. The term comprises a set of seventeen obscure elements with barely pronounceable names, like neodymium and yttrium. They are crucial ingredients in many high-tech products, including mobile phones, wind turbines, electric vehicles, and military weapons systems. At that point in 2010, Chinese mines supplied 95 percent of the world's rare earths. There was no alternative source to which

Japan's massive electronics industry could turn. A Chinese embargo could deal a heavy blow not just to Japan's economy, but to the whole Western world it was connected to.

Chinese officials never said outright that they were cutting Japan off; it just so happened, as David S. Abraham writes in *The Elements of Power*, that "all thirty-two of the country's exporters of rare earth elements halted trade on the same day." Panic seized the markets. Rare earth prices shot up as much as 2,000 percent. The sudden spike in costs bankrupted as many as fifty-nine renewable energy companies and unnerved business executives around the world. The United States, Japan, and the European Union filed a complaint against China with the World Trade Organization. But everyone recognized that the incident portended much more than a run-of-the-mill trade dispute.

Whether or not Beijing had directly ordered an embargo, political and business leaders in the West were suddenly jolted into realizing that China *could* cut any of them off at any time. Beijing had a stranglehold on a major choke point in the world's economy—one that, years earlier, the United States had essentially handed it. Now, American leaders were beginning to understand how vulnerable that choice had left the country. The House Armed Services Committee convened a hearing on the matter. "The world's reliance on Chinese rare earth materials, in combination with China's apparent willingness to use this reliance for leverage in wider international affairs, poses a potential threat to American economic and national security interests," declared then-Representative Ed Markey of Massachusetts.

With a metaphorical gun made of rare earths pointed at its head, Japan caved. Zhan was sent home on a chartered plane to his home province of Fujian, where he received a hero's welcome. Rare earths began flowing out of China again.

But the Western world recognized it could not go back to business as usual. In the wake of the brief embargo, the United States, Japan, and the European Union all began concerted efforts to find and develop sources of rare earths outside of China's control. The episode was one of the first major skirmishes in the escalating global struggle for critical metals.

Rare earths are actually neither rare nor earths. Most of them are quite abundant, but they are almost never found in their pure form. Instead, they come dissipated in very low concentrations within other minerals, like grains of pepper in a meatball. That makes them difficult and expensive to separate out. Many of the rare earth metals are often found clustered together. The seventeen rare earths, in atomic-number order, are scandium, yttrium, lanthanum, cerium, praseodymium, neodymium, promethium, samarium, europium, gadolinium, terbium, dysprosium, holmium, erbium, thulium, ytterbium, and lutetium. Rare earths are like vitamins, or spices, or yeast, or LSD: They are used in tiny amounts but have major effects. Each is unique, but, in general, they have unusual magnetic and electrical properties that can confer additional powers to other materials with which they are alloyed. "They enable both the hardware and the software of contemporary life to be lighter, faster, stronger, and longer ranging," writes Julie Michelle Klinger in *Rare Earth Frontiers*. "Global finance, the Internet, satellite surveillance, oil transport, jet engines, televisions, GPS, and emergency rooms could not function without rare earth elements."

The ill-fitting name dates back to the eighteenth century, when a Swedish artillery officer and amateur chemist found a very unusual rock in the village of of Ytterby, near Stockholm. A Finnish chemist named Johan Gadolin analyzed it and declared it a new type of "earth." Over time, scientists separated out the several elements in the rock. If their names sound made up, that's because they are. Yttrium was named after the village, gadolinium after Gadolin, scandium after Scandinavia. (Cut those chemists some slack. All the easy names, like gold and tin, were already taken.)

These new elements remained little more than curiosities for the next century or so, little-visited cubicles on the periodic table. Then, in the late 1800s, a German chemist figured out how to use one of them, cerium, to create the gas mantle—a fabric bag that lights up when heated. Billions of these mantles would eventually be used as streetlights. That was great for Germans who enjoyed an evening stroll but awful for people in the distant countries where the rare earths were mined. In those places, acidic water from the mines leached into local water supplies. "At the time, these costs were hidden to Western consumers, because they were happening in faraway places like Brazil, India, and South Africa," Lisa Berry Drago explained on the podcast *Distillations*.

By the second half of the twentieth century, America became the world's top producer of these curious metals thanks to a single mine. In 1949, prospectors in the high-desert scrublands of eastern California, while looking for uranium to build up America's nuclear arsenal, stumbled across a huge deposit of bastnaesite, an ore containing fifteen of the seventeen rare earths. They were disappointed to learn that the radiation they had detected came not from uranium but from thorium, a radioactive element that is often found with rare earths. The Molybdenum Corporation of America purchased the mining claims at the site, dubbed Mountain Pass, and quickly developed a lucrative business selling europium, a rare earth that enhances the ability of glass screens to display colors. There was a booming demand for europium from makers of the hottest new high-tech product of the day: color television. Scientists quickly found new uses for the neodymium, praseodymium, and other rare earths in the ore, especially in the growing consumer electronics industry.

For several decades, Mountain Pass was the world's leading source of rare earths. But in the 1990s, the Environmental Protection Agency discovered that a wastewater pipeline from the mine had ruptured dozens of times, "spraying the soil and surrounding vegetation with mineral slurry containing toxic concentrations of lead, uranium, barium, thorium and radium," according to Klinger. At least three hundred thousand gallons of radioactive waste spilled into the desert. More leached out of holding ponds and into the groundwater.

Facing huge fines and cleanup costs, the mine was closed down. The pit and its sprawling complex of support buildings sat for years under the Mojave Desert sun, silent and idle as a lizard. But it was not to stay that way.

By the time Mountain Pass closed, manufacturers were using rare earths for a dizzying range of obscure but important tasks in all kinds of products, from portable X-ray machines to camera lenses. Their use has only continued to grow since. Smartphones are jammed with rare earths: europium and gadolinium in the screens, lanthanum and praseodymium in their circuitry, terbium and dysprosium in their speakers, and more. Many military technologies used by the armed forces of the United States and other countries, including lasers, radar, night vision systems, missile guidance systems, jet engines, and alloys for armored vehicles, also depend on rare earths.

But the number-one product we use rare earths to make is permanent magnets—components that convert movement into electricity, and electricity back into movement. Scientists began developing permanent magnets in the 1980s, when they figured out that adding a little bit of rare earth metals like neodymium and dysprosium to common metals like iron and boron produced a very powerful magnet. Tiny versions of these magnets make your cell phone vibrate when you get a call. Bigger ones turn the wheels of electric cars. And really big ones in wind turbines convert the movement of air into electricity. A single wind turbine can require as much as five hundred pounds of rare earth metals.

In the 1990s, almost all of these magnets were made in the United States, Japan, and Europe. Ten years later, most were made in China. That didn't happen by accident. In the mid-1990s, two Chinese companies bought an American company called Magnequench, a GM spinoff that held the patent for a key type of neodymium-based magnet. The American government approved the deal on condition that the company stay in the United States for at least five years. "The day after the deal expired," writes Sophia Kalantzakos in *China and the Geopolitics of Rare Earths*, "the company shut down its US operations; employees were laid off, and the entire business was relocated to China."

The shift of the magnet industry to China was part of the epochal shift from West to East of so many heavy industries. This happened partly because developing countries—especially China, which began opening its economy to the global market in the 1970s—were eager to build their industrial bases, had lots of natural resources, and had millions of people willing to work for far lower wages than their Western counterparts. But it also happened because Western countries were tired of the pollution and destruction caused by industries like mining and started leaning on those industries to clean up their acts. "Things began to change in the 1970s," laments a *BusinessWeek* article from 1984 titled "The Death of Mining in America." "The North American industry was hit with large environmental expenditures. New competitors from the Third World appeared, many of them state-owned and blessed with abundant reserves."

It wasn't just the budding environmental movement that was pushing for a cleanup. Conservatives, too, felt a patriotic duty to keep America beautiful. It was President Richard Nixon, of all people, who created the Environmental Protection Agency in 1970. (Oddly, Nixon's brother Edward later founded a company that helped sell Chinese rare earth products to Western industries.) Many felt that offshoring and outsourcing production was in the nation's best interest. That way, Americans could just buy the products they needed, while the pollution generated by making those products stayed in someone else's backyard. "In the last two decades of the twentieth century, the workload of the future energy and digital transition fell naturally between China, which did the dirty work of manufacturing green-tech components, and the West, which could then buy the pristine product while flaunting its sound ecological practices," writes Guillaume Pitron in *The Rare Metals War*. But everything has a cost. America sloughed off its pollution onto other countries, but, in the process, it lost control over many of what have since become critical industries.

China saw the importance of rare earths early. The country holds as much as one third of all the world's rare earths, including perhaps the single largest deposit of the metals, at the Bayan Obo mine in the province of Inner Mongolia, northwest of Beijing. The Chinese government began investing heavily in the industry in the 1970s, declaring rare earths a "strategic mineral." Chinese officials made no secret of their ambitions. "Improve the development and application of rare earth, and change the resource advantage into economic superiority!" declared Chinese Communist Party general secretary Jiang Zemin in 1991. "The Middle East has oil. China has rare earths," Deng Xiaoping added in 1992.

Bayan Obo's ore is processed in the nearby city of Baotou, home to some two million people. Today, the area is the biggest hub of rare earth production in the world. Not by coincidence, it is also one of the most polluted areas on the planet.

Producing metals can generate pollution in many ways at each step of the process—mining the ore, refining it, and dealing with the waste by-products. Blasting ore out of the ground also blasts out toxic elements, which can be carried on the wind into rivers, farm soil, and people's lungs. Separating any metal from its surrounding host rock requires crushing, smelting, and often treatment with chemicals. Rare earths are particularly tricky, partly because they appear in such small concentrations, and partly because there are so many of them. It often takes dozens of steps to separate rare earths from the ore and then from each other, including baking the ore at extremely high temperatures and bathing it in acid. Rare earths are also often found mixed with radioactive materials, such as thorium and uranium, which have to be separated and somehow disposed of. These processes also often release toxic gases and require water, which gets contaminated. At Bayan Obo, writes Klinger, "Every ton of rare earth concentrate produced generates approximately one ton of radioactive wastewater; seventy-five cubic meters of acid wastewater; 9,600 to 12,000 cubic meters of waste gas containing radon, hydrofluoric acid, sulfur dioxide, and sulfuric acid; and approximately 8.5 kilograms of fluorine."

All of this can result in a huge range of poisons accumulating in the land and human bodies. By-products from mining and refining rare earth ores have polluted the nearby Yellow River and sown skeletal deformities and cancers among the inhabitants of the Baotou region. "Twenty minutes outside of Baotou," writes Aaron Perzanowski in *The Right to Repair*, "sits a toxic lake described by the BBC as a 'nightmarish . . . hell on earth.' It is filled with 'black, barely-liquid, toxic sludge'— the by-product of the nearby Baogang Steel and Rare Earth mine. This noxious muck has leached into local waterways and irrigation systems with devastating consequences. Decades before it became the center of the rare earth trade, Baotou was surrounded by fields of watermelons, eggplants, and tomatoes. These days, the soil can no longer support crops, the livestock has died off, and residents are battling leukemia and pancreatic cancer. Others report their hair and teeth falling out."

To American policymakers, China's rare earth industry is more than a threat to the local ecology—it's a threat to US national security. Remember, China mines most of the world's rare earth ore, processes almost all of it regardless of where it was mined, and manufactures the majority of the rare earth–powered permanent magnets that make windmills and electric cars work.

"China represents a significant and growing risk to the supply of materials and technologies deemed strategic and critical to US national security," declared a 2018 Department of Defense report, singling out rare earths and permanent magnets as particularly worrisome vulnerabilities. "When China needs to flex its soft-power muscles by embargoing rare earths, it does not hesitate." Nations cutting off supplies of important resources to their rivals is not a new tactic. OPEC member states imposed an oil embargo on Israel and its allies after the 1973 Yom Kippur War. The US itself banned exports of helium to Nazi Germany in the 1930s and of wheat to the Soviet Union after its invasion of Afghanistan in 1979. China is no different. In recent years, Beijing has threatened to throttle rare earth exports to American defense contractors in retaliation for US support for Taiwan. In 2023, China restricted exports of gallium and germanium, non–rare earth metals used for solar panels and electric car components, as well as graphite, a battery ingredient, in apparent response to Washington cutting off sales of computer-chip technology to some Chinese companies. "In this critical minerals and materials context we are up against a dominant supplier that is willing to weaponize market power for political gain," US energy secretary Jennifer Granholm said in late 2023.

So, it's little wonder that the United States, Western Europe, and their allies are, however belatedly, hunting for rare earth supplies in friendlier terrain.

CHAPTER 3

The Global Treasure Hunt

The roar of the explosion surges up like a reverse sonic waterfall from the bottom of the gigantic pit. I feel the ground shudder under my feet, even though I'm standing hundreds of yards away. The controlled blast has just shattered 150,000 tons of solid rock into a pile of boulders and loose stones. A cloud of brown dust and debris surges into the desert air. It's about as vivid a declaration of life as you could ask for from the old Mountain Pass mine in California, now back in action after years of lying dormant.

"That's a lot more exciting than watching a Bloomberg terminal all day," says Michael Rosenthal. We're standing on the lip of the pit on a cloudless, shirtsleeve-warm day in January of 2022.

A slender man in his early forties wearing a safety helmet decorated with a large mustache sticker, Rosenthal is the chief operating officer of MP Materials, the company that owns and operates the Mountain Pass mine. The site is hard to miss. You're driving down I-15 about an hour south of Las Vegas, surrounded by nothing but miles of desert stubbled with creosote bushes and the occasional Joshua tree. Then suddenly, on a ridge to the north, there's an eruption of hulking industrial machines and buildings. It looks like a small, fortified town of sandstone- and ochre-colored structures and towering white storage tanks connected by networks of pipes and conveyor belts. When you look closely, you realize that the hill nearest the site isn't a hill at all—it's an enormous pile of tan and black crushed rock, leftovers from ore dug out of the ground. The operation comprises over $1 billion worth of machinery and equipment spread over 2,200 acres.

Before he started at MP Materials, Rosenthal had no experience whatsoever with mining or any kind of heavy industry. It's something of an accident that he found himself in charge of a gigantic hard-rock mine. What he knows about is money. Rosenthal is an investor by training and trade, a man who makes his living looking for undervalued assets, buying them at a discount, and then selling them to someone else. The old Mountain Pass mine came to his attention in 2017.

A company called Molycorp Minerals had acquired the shuttered site in 2008. The new owners sank nearly $2 billion into upgrades and improvements but went bankrupt before they could figure out how to make the mine run profitably. "People in the finance world knew about Molycorp, because it was really hyped, and then it collapsed so quickly," Rosenthal says. "I felt like, 'Oh, wow. There's just gotta be something here. They put so much money into it.'" He knew how important rare earths were, and that many people wanted to develop supplies outside of China's control. Here was Mountain Pass, a rare earth mine that was once the world's leading producer, located right in the US of A, complete with all the environmental permits it needed. "It's got to be worth something," he figured.

Rosenthal and a partner put together a group of investors and bought the whole operation for several hundred million dollars—a deep discount on what Molycorp had sunk into the place. The plan was to flip it for a quick profit. The problem was, no one wanted to buy it.

Not without reason. The mine was bleeding what Rosenthal calls "an ungodly amount of money." It lost around $150 million the last year that Molycorp ran it. The consensus was that it was a lost cause. California's environmental regulations were too onerous, people said. And the Chinese could put it out of business at any time. "No major industrial company or financial company, at all, was even willing to sign a non-disclosure agreement to get more information about the site," says Rosenthal. "They were so convinced it was hopeless." The only way the investors could recoup their money was to take over the business themselves and find a way to turn it around. So, they hired a crew of mining professionals, including many who had worked at Mountain Pass under Molycorp, and got the place running again.

A recently bankrupted mine that that no one wanted to buy run by two businessmen with no mining experience doesn't sound like a sure formula for success. Nonetheless, Mountain Pass has done surprisingly well. In its first year of operation under the new owners, the mine produced about fourteen thousand tons of rare earths. In 2022, that number topped forty-two thousand tons—more than triple what Molycorp ever managed in a single year. That's nearly 15 percent of all production worldwide. (Though the quantities of rare earths extracted globally each year are growing fast, they are still tiny compared to other metals. About three hundred thousand tons of rare earths were extracted in 2022, a fraction of the millions of tons of copper, let alone the billions of tons of iron mined each year.) MP Materials pulled in about $527 million in revenue in 2022. It is now by far the biggest rare earth producer in the western hemisphere.

That success is due, at least partly, to outside help. The company has benefited handsomely from the US government's push to build up the domestic critical-metals industry. The company isn't shy about pitching itself as an explicitly American asset. "We believe we can generate positive outcomes for US national security and industry, the US workforce, and the environment," the company proclaimed in a 2021 corporate SEC filing. The Department of Defense has pledged $45 million, spread over several years, to help scale up the Mountain Pass operation.

Mountain Pass's pit is small by the standards of major mines, but it's still impressive. It's the equivalent of forty stories deep and as wide as a canyon, its bottom accessed by a rough road that switchbacks down its sides. The giant trucks and front-end loaders at the bottom look like tiny toys from the pit's edge. And this pit will just keep getting bigger and deeper for the next twenty years, one dynamited chunk at a time, if the mine continues operating as planned.

Explosions in the pit bottom, like the one I saw, are one of the early steps in the long and convoluted process of producing rare earths, and for that matter most other critical metals. Rare earths tend to come grouped together, with several present in the host rock—Mountain Pass's bast-naesite ore contains fifteen of the seventeen rare earths, all but scandium and promethium—but they make up only a small fraction of that rock, and the concentrations vary widely. Mountain Pass's rock is 7 to 8 percent rare earths, on average. That's a very high concentration by industry standards and one of the reasons the mine is so promising. Still, separating the valuable rare earths from the surrounding junk rock takes a lot of muscle, as well as a lot of finesse.

The process of converting raw, earthbound rock into pure metals often starts with blowing it up. The American mining industry alone uses well over one million tons of explosives each year. To prepare for the blast I saw, engineers examined a roughly rectangular outcropping of solid rock at the pit's bottom and identified its fissures and weak points. They then drilled 236 holes into the rock, each about thirty-four feet deep, packed them with ammonium nitrate, and detonated all of them simultaneously. When the dust cleared, the rock slab was . . . still there, more or less. Only now, instead of one solid, flat-topped mass, it had been shattered into boulders and loose stones that were stacked together like a bunch of misshapen building blocks.

All those drill holes were carefully calibrated to break up the rock without sending it flying off in all directions. That makes it much easier to haul the ore up out of the pit. "The goal is to have as little horizontal motion as possible. You just kind of shock it into cracking," explains

Rosenthal. "Everything should basically be exactly where it was, just now loose."

Earth-moving machines then load the loose rock into hundred-ton dump trucks, which haul it out of the pit to the surface and over to a crushing machine. In the crusher's maw, the boulders are shattered into pebbles small enough to fit in a thimble. A dusty conveyor belt carries the pebbles up some thirty feet high, then dumps them into a huge, ever-shifting pile.

Those pebbles are then trucked down to a large building crammed with pipes, ducts, walkways, and holding tanks. Inside, the building reverberates with a multilayered industrial cacophony of motors, conveyor belts, and clangor from the key piece of gear: the spinning ball mill. The ball mill is a studded metal cylinder as long as a semitrailer truck filled with hundreds of two-inch steel balls. Inside the mill, the pebbles are mixed with water and spun around and around, the steel balls bashing and smashing them down into a powdered slurry.

The next step is to separate out and concentrate the grains of powder that contain bastnaesite—the ore that contains the rare earths—from the grains that are commercially worthless mineral junk. The slurry is run through a series of tanks in which it is treated with chemicals that make the bastnaesite hydrophobic—literally translated: afraid of water. Air is piped through the mixture, and the particles of bastnaesite, desperate to escape the terrifying water around them, latch on to the air bubbles, which carry them to the top of the tank, where they emerge amid a thick, bubbling goop. Heavy mechanical presses then squeeze the water from the goop. What's left after that first separation phase is a powder that is approximately 80 to 90 percent pure bastnaesite. Also left over: boatloads of wastewater.

Mines of all sorts all over the world use similar procedures to process the stone they dig up. Handling the resulting wastewater is a major issue. Typically, mining companies dump wastewater into what are known as tailings ponds—artificial lakes loaded with crushed rock and chemicals left over from this process. Over time, if all goes well, the solids settle to the bottom of the pond, and the water evaporates or is recycled. These ponds are supposed to be engineered to keep their toxic contents safely away from everything else, but there's a constant risk that polluted water will seep out of the pond and into the local groundwater. Toxic wastewater leakage is what got the Mountain Pass mine shut down in the first place.

There's also the risk that the dams holding back the ponds will give way, unleashing catastrophic mudslides. The worst such dam failure in US history (so far) claimed 125 lives at a West Virginia mine in 1972. That was eclipsed in 2019 by a mining-dam collapse in Brazil that unleashed a torrent of toxic sludge that polluted nearly two hundred miles of rivers and killed 270 people.

The new version of the Mountain Pass mine, however, is one of a few mines around the world that has no tailings ponds. The slurry is run through a giant filter that separates the solid tailings from the water. The tailings are then dumped in a lined pit, which will eventually be sealed and buried. The water, meanwhile, is recycled within the plant, providing most of the facility's needs. (The rest of its water comes from wells the company owns in the nearby desert.)

Even after all this, there's still a long way to go. All the rare earth elements are still mixed together within the bastnaesite powder produced in the first processing phase. Many more steps are required to separate them out. But, at the time of my visit, MP Materials wasn't able to perform any of those steps. Instead, they packaged their bastnaesite in giant bags and sold it to another company . . . in China. That company, Shenghe Resources, owns just shy of 8 percent of MP Materials. "We have to export it," says Rosenthal. "There's no processing facilities anywhere outside of China that can handle the scale that we need to be producing."

MP Materials is building further refining capabilities to its Mountain Pass operation and aims to open a rare earth–separation and –refining plant in Texas. The ultimate goal is to build an entire US-based rare earth–magnet supply chain, one that can compete with China at every stage, from mine to manufacturing. General Motors has committed to buying alloy and magnets when (and if) MP starts making them. Lynas Rare Earths, an Australian company that is currently the world's top non-Chinese producer of refined rare earths, is also building a plant in Texas, with more than $150 million in support from the US Department of Defense. But even if both projects come online this decade, as planned, they will still only be able to supply a fraction of the world's demand.

Meanwhile, a flock of other players are scouring the world for commercially viable new deposits. But, even in this era of satellite imaging and ground-penetrating radar, finding significant amounts of any critical metal is startlingly difficult. Most of the world's easily discovered reserves are already being tapped. The ones that remain tend to be in remote locales and deep underground. Miners generally say only one in one hundred exploratory boreholes turns up anything.

Prospecting works a bit like the tech industry. Much of the riskiest legwork—the actual prospecting—is handled by startups, small companies dubbed "juniors." These can comprise just a few people with a hunch and a staked piece of remote land. Like tech startups, juniors take on the risk of developing unproven new ideas. Most fail and disappear, but a few work out well enough to be bought by a more established player.

"These small crews get someone to give them a couple of million bucks," explains one industry veteran. "They have no assets, and their shares are at two cents. If they find huge resources, everyone is suddenly rich. But the vast majority burn through that money and find nothing. Or they find something but not enough of it. Or the market changes, and that thing isn't in demand anymore. So, the big companies just wait for juniors to find something, then buy them."

Take Steve Mynott, a young man with neatly gelled hair and a pink shirt under his gray suit, whom I met in an office shared by several tiny mining outfits in Vancouver, Canada. He was aiming for a career in professional hockey until an accident at the age of eighteen broke his back and killed his dream. He drifted around for a while until a skating buddy found him a job cold-calling potential investors in a nickel mine. "I knew nothing about it, but I like talking to people. Fifteen years later, I'm still here," says Mynott. When I met him in June of 2022, he was the CEO of Eagle Bay Resources, a junior that owns some land in British Columbia that Mynott is pretty sure contains a profitable rare earth deposit. Proving that, however, is difficult. The land is deep in British Columbia's remote, mountainous interior. "It's rocky terrain, most of it steep, and there are bears all over the place," he says. Mynott had raised $1 million and spent most of it acquiring the property and sending geologists out there to gather soil and rock samples. The next step was to do some exploratory drilling, which required yet more funding. Mynott's main task, at that point, was trying to scrounge up the necessary cash. "It's a lot of picking up the phone, hustling, grinding to get money," he says.

A handful of startups are trying to make prospecting faster, cheaper, and more efficient by applying artificial intelligence. California-based KoBold Metals has built a gargantuan database incorporating all the information it can find about the Earth's crust—the equivalent of thirty million pages of geologic reports, soil samples, satellite imagery, academic research papers, and century-old handwritten field reports. All of that gets run through artificial intelligence algorithms that identify patterns in the geology and other features of places where metals have been found in the past. The algorithms can then be turned loose on the full database to find similar patterns at locations that haven't been explored, spitting out a series of maps indicating where the target metals are likely to be found.

KoBold has sent prospecting teams to areas in Zambia, Greenland, and Canada. "We're looking to expand and diversify the supply of critical metals all over the world, but we're taking a totally different approach" from conventional mining companies, KoBold cofounder Kurt House says. "Two thirds of our team are software engineers or data scientists who have never worked a day of exploration in their life. The other third are experienced explorers."

While most AI exploration companies sell their services to mining outfits, KoBold aims to take part in the actual extraction operations. It currently holds the exploration rights to thousands of square miles of land all over the world and has struck deals with some of the world's largest mining companies, including BHP and Rio Tinto.

Even with help from AI, placing bets on potential mineral deposits is still very risky; metals often turn up in places with wildly different conditions and geologic histories. "When you're training an algorithm to recognize a face, you can assume there's a mouth and it's below the nose and eyes," says Sam Cantor, head of product at Minerva Intelligence, another AI-driven mining-exploration startup. "But if you apply that training to insect faces, you might find more than two eyes and no nose. Training an algorithm on data from Alaska and applying it to Nevada means it might have a lot of wrong assumptions." KoBold has generated a lot of buzz and investment from the likes of venture firm Andreessen Horowitz and Bill Gates's Breakthrough Energy Ventures. But, five years after the company launched, it had yet to get past the exploratory-drilling phase.

New rare earth mines have been opened in recent years in Canada and especially in Australia, where Lynas controls one of the world's biggest deposits. Others, however, have been blocked by locals unwilling to shoulder the environmental burden that mines inevitably bring. Greenland holds enormous troves of rare earths, but, in 2021, angry locals and legislators thwarted efforts to open a major proposed mine. Protesters, including Greta Thunberg, similarly stalled another potential mine in Sweden. And Lynas's rare earth processing plant in Malaysia may be forced to close in the coming years under pressure from government regulators.

Even China has grown tired of the havoc wrought by rare earth mining. It has maintained its grip on the less-polluting, higher-value refining and processing stages, but it is drawing back from the dirty business of extracting ore from its own territory. Starting around 2016, the Chinese government began cracking down on illegal miners, consolidating the big rare earth manufacturers, and closing down many of its most polluting mines. But the country's rare earth refineries still need raw materials to process and sell to the world. So, it has followed the example that the United States set in the twentieth century. Today, China outsources much of the destructive, polluting work of mining to the neighboring nation of Myanmar.

Myanmar is especially rich in so-called "heavy" rare earths, including dysprosium and terbium, which are critical in the production of permanent magnets. In 2014, Myanmar exported just $1.5 million worth of rare earths to China, according to a deeply researched report by Global Witness, a UK-based nonprofit that monitors extractive industries. By 2021, that figure had rocketed to $780 million, and it has continued to climb since. That ranks Myanmar among the world's top producers of rare earths. It now supplies nearly half of China's heavy rare earths.

Those metals come at a bitter cost. They are extracted mainly in a secluded, mountainous corner of the country controlled by armed militias connected to Myanmar's brutal military regime. Often, the militias have no legal rights to the land they occupy—they simply seize it from the Indigenous Kachin people. "No one wants to give up their ancestors' lands, but if they [resist], they can be killed," a local resident told Global Witness. "Rare earths from Myanmar should absolutely be treated as conflict minerals. They're being mined in conditions of armed conflict, and they are sold by armed groups," Clare Hammond, a campaigner with Global Witness, told *Myanmar Now* in 2021.

The process of extracting those metals is gruesome. Miners drill holes into mountainsides and inject them with an ammonium sulfate solution that liquefies the earth. The mix of chemicals and soil percolates down through the mountain and drains out into collection pools, from which the minerals are leached out. Global Witness commissioned a satellite to fly over the region in 2022 and spotted more than 2,700 of these bright blue, chemical-saturated pools.

Myanmar's rare earth industry has caused landslides and spread poison into the country's soil and rivers, occasionally turning them red with waste, according to Global Witness and reports by the Associated Press, local media, and other sources. Animals—including endangered tigers, pangolins, and red pandas—have abandoned the area. "In the areas we worked on in the mountains, there's nothing left, not even a small bird," one miner told the newspaper *Frontier Myanmar*.

All of these rare earths are sold to China, where they are refined and sold to the world. In other words, there's an excellent chance the electric car you drive or the cell phone you text on might include metal unearthed at the cost of poisoned waterways, dead animals, and terrorized civilians in Myanmar. Everything has a cost. But not everyone pays it.

What's more, Myanmar is not the only place where the critical metals making the energy transition possible are being gathered at gunpoint.

CHAPTER 4

Killing for Copper

Moqadi Mokoena had been feeling uneasy all day. Earlier, when he left his home on the outskirts of Johannesburg, South Africa, for his job as a security guard, he had to turn around and come back twice, having forgotten first his watch and then his cigarettes. He had reason to be nervous. His supervisor had assigned him to join a squad protecting an electrical substation where, two days earlier, four other guards had been stripped naked and beaten with pipes by gun-wielding thieves. Now, on this day in May of 2021, Mokoena and a fellow guard were at that same substation, peering tensely through their truck's windshield as a group of armed men approached.

Mokoena pulled out his phone and called his wife, Itumeleng, the mother of their one-year-old daughter. He told her about the gang coming towards him. "I'm feeling scared," he said. He didn't have a gun himself. "I think they are the same ones who attacked our colleagues."

"Call your supervisor!" she told him.

Minutes later, the men opened fire with at least one automatic weapon. Mokoena's partner jumped out of the vehicle but was cut down by bullets. A third guard nearby dove for cover, shot back at the thieves, then ran for help. When he returned with his supervisor, they found Mokoena and his partner dead.

"We face these dangers every day," the surviving guard later told a local journalist. "You don't know if you'll return home when you leave for duty. It's scary, but I need to put food on my table."

In most places, electric-power companies are pretty dull businesses. But in South Africa, they are under a literal assault, targeted by heavily armed gangs that have crippled the nation's energy infrastructure and claimed an ever-growing number of lives. Practically every day, homes across the country are plunged into darkness, train lines shut down, water supplies cut off, and hospitals forced to close, all for lack of electricity. It is disruption on a scale usually seen only in wartime. But the attackers in this conflict aren't trying to seize territory. They're not even trying to disrupt civilian life. They just want copper.

The battle cry of energy-transition advocates is "electrify everything." Meaning: Let's power cars, heating systems, industrial plants, and every other type of machine with electricity rather than by burning fossil fuels. It's the right idea, but it raises an often-overlooked but enormously important point: While we need to build renewable-electricity sources, we also need to massively upgrade the electric grids in practically every country. We will have to strengthen and expand the electric infrastructure everywhere to connect all those new power sources with all the new charging stations we'll have to build to fuel all the new electric cars. In the United States alone, the capacity of the electric grid will need to grow as much as threefold to meet the demand.

Electric grids are mainly composed of miles and miles of copper cables and wires. Copper is the metal that carries the electricity from wind turbines in New Mexico to my garage in Los Angeles. It is literally the medium of power. There's no beating copper's usefulness in this arena. It's highly ductile, meaning it can be shaped and pulled into wires without breaking. Even better, copper conducts electricity more efficiently than any other metal except silver, which is too expensive to use on such a scale. No wonder Goldman Sachs has declared "no decarbonization without copper" and called copper "the new oil."

Of all the metals critical to the energy transition, copper is the one we need the most of in sheer tonnage. The red metal not only composes the arteries of the electric grid, it is also a critical component of renewable-energy tech itself. Copper connects the cells inside solar panels, is used in the generators in wind turbines, and composes much of the battery, motor, and wiring systems in electric cars (a typical EV contains as much as one hundred seventy-five pounds of copper). The giant storage batteries that will be needed to store and provide power when the sun isn't shining and the wind isn't blowing will also rely on copper. About 60 percent of all copper produced each year winds up in electrical equipment.

No surprise, then, that demand for copper is surging. A recent report from market analysts S&P Global predicts that, by 2035, global demand for copper will more than double, from twenty-five million metric tons per year worldwide to fifty million tons, and it will keep growing from there. By 2050, annual global demand will add up to more than all the copper consumed on Earth between 1900 and 2021. "The world has never produced anywhere close to this much copper in such a short time frame," the report notes. The world might not be up to the challenge. Analysts predict supplies will fall short by millions of tons in the coming years. "Unless massive new supply comes online in a timely way, the goal of Net-Zero Emissions by 2050 will be short circuited and remain out of reach," the S&P report warns.

As the energy transition has gathered speed, requiring more and more electrical capacity, the value of copper has soared. Between March 2019 and March 2022, the price of a ton of copper shot from about $6,400 to more than $10,000. That has made electrical wiring, equipment, and even raw metal fresh from the mines into a juicy target for thieves. As the world's hunger for copper grows, so does a black market to feed it. In recent years, all around the world, including in the United States and Canada, hundreds of millions of dollars' worth of the metal has been stolen—and countless lives have been lost in the process.

With the possible exception of gold, no other metal has caused as much death and destruction. Most of the world's copper is produced and sold by legitimate companies, but even the legal copper industry has inflicted tremendous harm. From the western United States to South America to Central Africa, copper mining has left colossal pits full of toxic waste and fouled enormous swathes of land and waterways. In 2014, a giant copper-tailings spill in British Columbia annihilated a creek full of fish and the trees around it. In Mexico that same year, a leak of acidic copper sulfate contaminated the drinking-water supply of twenty thousand people. An earlier spill in the Philippines inundated villages, smothered coral reefs, and wiped out fisheries. A gas leak at a mine in Zambia in 2019 put more than two hundred schoolchildren in the hospital. The list goes on.

The risk of disaster is increasing, because most of the world's richest and most easily accessed copper deposits have by now been mined. "All the low-hanging fruit has been picked," Scott Dunbar, a Canadian former mining engineer, says. The quality of the remaining ore in many major mines—that is, the percentage of metal within the rock—is falling fast. The average grade of copper in Chile, the world's leading producer, has dropped to below 1 percent in the last fifteen years. A century ago, copper grades often topped 5 percent. That means ever-larger tracts of land have to be torn up to extract the same amount of copper, generating ever-larger amounts of waste. It also means more heavy machinery digging away for longer periods, using more energy, and generating more carbon emissions.

All mining causes some environmental damage, but copper stands out for the violence that often accompanies it. In Peru, police have killed protestors in clashes over copper mines. In Pakistan, struggles over copper resources are fueling an armed separatist movement. In southeast

Asia, copper provoked a full-scale war. The Panguna copper and gold mine on Bougainville Island, Papua New Guinea, was, for a time, one of the world's biggest, generating huge profits for its owner, Rio Tinto. Its mine tailings were dumped straight into a local river. In 1989, local residents, infuriated by the environmental damage and their meager share of the mine's profits, blew up the power lines feeding Panguna, forcing it to shut down. The central government sent troops in to restore order. The locals fought back, kicking off a civil war in which as many as twenty thousand people died. The shooting has stopped by now, but communities near the mine say that the more than one billion tons of mine waste dumped into the local river delta has left them with poisoned water and polluted farm fields.

At the same time, copper has also brought us tremendous benefits—perhaps more than any other metal. The telephone was made possible by copper. Train lines run thanks to copper. Hospitals control infections with the help of copper. And most important, electricity—the lifeblood of modernity, the elemental force that lights your home, powers your computer, charges your cell phone, and runs your air conditioner—is literally brought to you by copper. Whether that electricity was generated at a solar farm, a coal plant, a nuclear reactor, or a dam, the pathways that transport it are made with copper.

The red metal has been our ally from the dawn of human civilization and across many cultures. (Its English name comes via ancient Rome, where copper was called cyprium after Cyprus, the Mediterranean island that was the Roman Empire's main source. The word atrophied over time into cuprum and, eventually, copper.) Primitive peoples probably latched on to it because, unlike most metals, it can often be found in relatively pure form, is easily molded and shaped, and has a pretty, rosy color. Many cultures began using it to create simple weapons, tools, and adornments long before recorded history. In fact, using the metal helped create history. "The making of copper tools initiated a spectacular growth in human technology, being instrumental in the birth of other technologies, cities, and the first great civilizations," writes Mark Miodownik in *Stuff Matters*. The stone blocks that form the pyramids of ancient Egypt were carved with copper chisels. Archaeologists have unearthed a copper pendant more than ten thousand years old in northern Iraq, and a seven-thousand-year-old copper awl in Israel. The Bible gives a good sense of how treasured copper was: In Deuteronomy, God promises to lead the Israelites to "a land where bread will not be scarce and you will lack nothing; a land where the rocks are iron and you can dig copper out of the hills."

Ancient peoples in South America, sub-Saharan Africa, and the Indian subcontinent were also well acquainted with the red metal, using it to fashion everything from religious statues to cookware. In what is now called North America, Indigenous people were digging up copper in the Great Lakes region as long ago as 5500 BCE. The Haida people of western Canada considered copper more valuable than gold; a copper shield was the ultimate symbol of wealth.

Beginning somewhere around 3300 BCE, people in several places, including the Middle East, Europe, and what is now Thailand, figured out that melting copper and mixing it with a little tin yielded bronze, a far stronger and more durable metal that made excellent tools and even better weapons. That new alloy became so important that the subsequent two millennia or so are known as the Bronze Age. Bronze arrowheads, spear tips, swords, and armor became the must-have items for any serious army. Centuries later, metallurgists learned to alloy copper with another metal, zinc, to create brass, which we have come to use for musical instruments, coins, and many other practical items.

Over the centuries, we've found countless ways to use copper to improve our lives. Early printing-press plates were made from copper. Distillers of brandy, cognac, and fine whiskeys still use copper equipment to make their delightful beverages. Copper kills bacteria and has been used as a medical aid for centuries. In the late 1700s, Britain's Royal Navy figured out how to sheathe its warships in copper to keep them from rotting. Their rebellious American colonies copied the idea; Paul Revere, he of the famous midnight ride, was the first to manufacture sheathing for US warships.

It was thanks to its role in the epochal transformation of the Industrial Revolution that copper went from handy helper to indispensable pillar of civilization. In the 1830s, Samuel Morse developed the first models of his version of the telegraph, which sent messages in his eponymous code via electricity transmitted along copper wire. It was the dawn of the era of instant communications, and it came on fast. By the 1850s, the United States, Canada, and much of Europe had set up national telegraph systems based on copper wire. Telephones and electric power soon followed, both of which also required vast networks of copper cable.

While some inventors were devising ways to use copper to help people talk to each other, others were using it to help people kill each other. French armorers came up with the idea of encasing soft lead bullets in copper alloys, like brass, so that the bullets would better keep their shape when fired. The technique spread quickly. Demand for copper for munitions rose rapidly, and the insatiable needs of the armies fighting the First World War sent it into overdrive. Global production tripled from 1900 to 1918, to around 1.5 million tons per year.

Where was all that copper coming from? The lion's share was dug out of immense mines that opened up all over the western United States, particularly Arizona, Utah, and Montana. Meyer Guggenheim, patriarch of the famous family dynasty, made his fortune with copper mines in this era. In 1882, several outfits that eventually became the Anaconda Copper Mining Company started work on a phenomenally rich vein in Butte, Montana. By century's end, that area was dubbed "the richest hill on Earth" and supplied half of all of America's copper.

Montana's "Copper Kings" grew wealthy off the work of thousands of immigrants who streamed in to work the mines. Anaconda was not eager to share the wealth. "A ruthless (perhaps even murderous) opponent

of the mine workers' long struggle to unionize, an unapologetic manip-
ulator of Montana politics, and a censorious master of much of the state's
media, Anaconda ran Montana like a corporate fiefdom for a good part
of the twentieth century," writes Timothy J. LeCain in *Mass Destruction:
The Men and Giant Mines That Wired America and Scarred the Planet*. Violent strikes broke out regularly over the mines' brutal conditions. Fires,
falling rocks, and ill-planned dynamiting killed hundreds of workers
each year. Thousands more perished from diseases caused by breathing
in silicate dust. In 1917, Butte was the site of the worst mining disaster in
American history. A fire broke out in the shaft of one of the mines.
Smoke and poisonous gases billowed through the narrow underground
passages, suffocating men as they clawed at concrete barriers blocking
their escape. All told, 168 lives were lost.

The land suffered as well, mauled by both the colossal mines themselves and the pollution created in processing the rock that came out of
them. "The environmental effects of getting the copper out of the ore
were often even greater than those of getting the ore out of the mine,"
writes LeCain. Waste rock and tailings piles leaked acid and toxic metals
into rivers and streams. Smoke from smelters besmirched the air. Ranchers near the Butte mines saw their cows dying and sued Anaconda. The
company won.

All of this environmental degradation was perfectly legal. "Early
miners behaved as they did because the government required almost
nothing of them," writes Jared Diamond in *Collapse*. "Not until 1971 did
the state of Montana pass a law requiring mining companies to clean up
their property when their mine closed." The hole left behind when the
Butte mines shut down is now a 600-acre pit filled with 6.5 trillion gallons of poisonous sludge, thick with lead, cadmium, arsenic, and sulfuric acid. Flocks of unlucky geese have landed in it at least twice in recent
years; thousands of them died as a result. The pit sits right next to the city
of Butte.

Small wonder that Americans lost their appetite for copper mines.

Few tears were shed as the industry increasingly shifted overseas. Today, the United States produces only about 6 percent of the world's raw copper, ranking behind Chile, Peru, and China. The bulk of American copper now comes from Arizona, but, even there, the locals are getting fed up. In another example of how much influence mining's opponents have gained, environmental groups and local Apaches have stalled, for years, the potentially gigantic Oak Flat mine near Phoenix. Proposed mines in Alaska and Minnesota are similarly stymied.

The new frontier for copper is Central Africa. Investment has been pouring in since the early 2000s in answer to the howling need for raw materials from China's growing economy. A savage series of internal conflicts in several African nations had kept most outsiders away, but, by the 2000s, things had settled down, and eager capitalists rushed in from around the world. "For decades, the bulk of the continent had been neglected by big Western companies as too remote, too underdeveloped and too corrupt. Now they were falling over one another to invest," write Javier Blas and Jack Farchy in *The World for Sale*, a history of the modern commodity-trading industry. Most of the action is in the Democratic Republic of the Congo, a vast nation two thirds the size of all of Western Europe that is rich with copper as well as diamonds, gold, cobalt, and other minerals. Mining supplies around 80 percent of the whole country's foreign earnings. Little of that filters down to its people, however. Most Congolese subsist on less than $3 a day.

Early on the scene in the DRC was one of modern mining's most idiosyncratic moguls, Chicago-born billionaire Robert Friedland. At Oregon's Reed College in the early 1970s, Friedland became close friends with a younger but like-minded fellow student named Steve Jobs. Yes, *that* Steve Jobs. Both of them were entranced by Eastern religions and spiritual traditions. "On Sunday evenings Jobs and Friedland would go to the Hare Krishna temple on the western edge of Portland," writes Walter Isaacson in his biography of Jobs. "They would dance and sing songs at the top of their lungs." They both traveled to India to spend time with a famous Hindu guru. They also shared a fondness for hallucinogenic drugs. In fact, when he arrived at Reed, Friedland was on parole following a conviction for possession of twenty-four thousand hits of LSD.

In 1974, Friedland moved to an apple farm his uncle owned near Portland. "The farm became a hippy commune where groups of young people from the Hare Krishna temple worked the apple orchard, meditated, and ate vegetarian food together. Friedland took the name of Sita Ram Dass and looked 'like a Caucasian Krishna,'" recounts Henry Sanderson in *Volt Rush.* "Jobs, who had dropped out of Reed, worked at the orchard, helping to produce cider. Pretty soon, however, Friedland started operating the commune as a business, according to Jobs." The man who would later give the world the iPhone soon left, disenchanted by Friedland's "materialistic" impulses.

Friedland had decided (as Jobs later would) that making money wasn't so terrible after all. By the late 1970s, he had teamed up with some Vancouver-based financiers and moved into the world of mining, hustling for small gold outfits. He made headlines in 1992, when a Colorado gold mine he was involved with leaked poison into a river, earning him the nickname Toxic Bob. In the meantime, he discovered a major gold deposit in Alaska and an even bigger nickel deposit in Canada, both of which he later sold for billions of dollars.

Friedland has been a major player in the industry ever since. (He also has a sideline in movies, helping to produce *Crazy Rich Asians* and other films.) He has been involved in the DRC since 1996, when he met Laurent Kabila, then the leader of a rebel movement. "Friedland went on TV to defend Kabila's advance to power, and in return got 14,000 square kilometers of land outside the copper mining town of Kolwezi," writes Sanderson. The area sits in the heart of a phenomenally rich mineral belt stretching across the DRC and south into Zambia. Locals had mined it for centuries, and European colonists had done so for decades, but the civil war shut down most production until the early 2000s. Friedland's Ivanhoe Mines company found an enormous, high-grade copper deposit in the Kamoa area near the DRC's border with Zambia and, in 2021, started production at what is expected to become a colossal new mine.

That mine is a joint venture with a Chinese company, one of many that are heavily involved throughout the DRC's mining industry. China refines and consumes far more copper than any other country, and it relies on foreign suppliers for almost all of it. So, when the DRC began opening up to foreign investment, China pounced. Beijing has since supplied billions of dollars in infrastructure support to the DRC in exchange for far-ranging mining rights.

Friedland is a peerless pitchman for copper. "We simply cannot continue to exist as a species without a lot more copper, especially if we want to reduce hydrocarbon consumption, or if we want to electrify the world's transportation fleet," he told an industry conference in Santiago, Chile, in 2022. "The American electrical grid," he continued, "is a hodgepodge of poorly maintained ancient systems. It is not a modern electrical grid as the Chinese are constructing. It's a very old, very tired piece of junk." He continued: "Wind takes copper. The modernization of the grid takes copper.... Anything you want to do to get away from burning coal, anything you want to do to get away from burning hydrocarbon leads you directly to copper. There is no alternative." Copper from Friedland's new Kamoa mine will be some of the cleanest ever produced, he promises.

The world's number-one producer of raw copper, however, is on the other side of the world from the DRC. The elongated, underpopulated South American nation of Chile has held that title for decades. Chile contains the world's biggest reserves of the red metal, and it supplies nearly one quarter of all the world's raw copper. It is America's top foreign supplier. Copper has brought enormous wealth to Chile—as well as despoliation, dispossession, and violence.

Large-scale mining in Chile began more than a century ago, when Anaconda, flush with cash from its Montana base, expanded overseas. Among other acquisitions, the company took control of a site in northern Chile called Chuquicamata, where Indigenous people had been digging up copper for centuries. For a time, it was the world's most enormous open-pit mine. The path there was not a pretty one. Working conditions were grueling and strikes frequent. Miners' wives waited in long lines for whatever provisions the company store provided, and their husbands' employment category determined which company schools their kids could attend.

Among those struck by the harshness of the place was Ernesto "Che" Guevara, who visited Chuquicamata in March 1952. "One would do well not to forget the lesson taught by the graveyards of the mines, containing only a small share of the immense number of people devoured by cave-ins, silica and the hellish climate of the mountain," he wrote in his *Motorcycle Diaries*. When Chileans elected socialist Salvador Allende as president in 1970, one of his first moves was to nationalize the country's copper mines, including Chuquicamata. The mines' American owners were furious. Three years later, Allende was overthrown in a US-supported military coup.

Today, Chuquicamata is run by Chile's state-owned copper company, Codelco. The mine complex sprawls along the lower reaches of a range of hills in the Atacama Desert in Chile's far north. It's one of the driest regions on the planet, and it looks the part—miles of sand and rock with barely a speck of vegetation, bounded in the distance by the snow-capped Andes. The pit itself is gargantuan: more than a mile wide, two miles long, and most of a mile deep. It's the center of a sprawling complex of buildings and roads teeming with house-sized trucks, hulking machinery, and towering chimneys. It squats on the hillside swathed in dust and smoke like an industrial suburb of Mordor.

But the mine's impact on the surrounding landscape is even greater, reverberating out like shock waves from an enormous bomb blast. In every direction, the land has been scarred, rendered a literal wasteland, all in the service of the mine. Mountainous heaps of tailings, candy-striped with layers of gray and brown minerals, sprawl away for miles.

They could easily be mistaken for hills, matching as they do the size of the naturally occurring ones around them, except for their flat tops and overly tidy shapes. The trucks ferrying waste rock to them look like insects in comparison. Electricity towers march across the landscape, bringing in power like soldiers bringing provisions. A nearby settling pond spreads over an area the size of Manhattan, its bottom lined with thick black plastic, its polluted water held in by berms of bulldozed earth and rock.

And then there's the damage you can't see. The mine, and others nearby, draw huge amounts of water from the aquifers and streams of the Atacama Desert. Not far from Chuqicamata is the Escondida copper mine, currently the world's biggest. It was sanctioned by the Chilean government for overdrawing water and, as a result, now has to pump in desalinated seawater up ten thousand feet and over one hundred miles from the Pacific Ocean.

Leonel Salinas, president of the village of Lasana, a tiny collection of small houses along the Loa River about thirty miles from Chuquicamata, says mining is sucking away what little water the area has. Farmland is drying up, forcing the mostly Indigenous locals to move away to the cities. "We're losing our cultural landscape," he says. "We're not against modernity and development, but the burdens need to be equitably shared."

Dina Panire, president of the nearby farming town of Chiu Chiu, feels the same. "We've been hurt for years by the mines," she says. "There used to beautiful springs and wetlands here. They've disappeared. The other regions get the money, but we here are the ones who pay the price."

The treasures these mines produce attract some astonishingly brazen criminals. Thieves have stolen copper slabs from industrial plants and wires from electric and communications utilities, causing thousands of service outages. But for cinematic flair, there's no topping Chile's train robbers. By the light of the full moon, bandits in Toyota

Tundra pickups roll up alongside trains that are hauling copper slabs from the mines high in the Atacama down to the coast. With perhaps a whispered prayer to the spirits of Butch Cassidy and the Sundance Kid, the bandits leap aboard the train cars, slice through the ropes securing the 180-pound slabs, toss them into the beds of the speeding trucks, and disappear into the night.

The problem is so acute that the Chilean national police have set up a special copper task force. But thieves were still robbing trains regularly when I visited Chile in 2022. And not just trains, for that matter. In January of 2023, Chile was hit with one of the most brazen heists in recent memory. Ten men driving trucks cut the security cameras at the country's main seaport, overpowered a handful of workers, and made off with a dozen containers full of Codelco's copper—more than $4 million worth.

Other critical metals also get stolen, of course, and not just in Chile. US authorities have shut down at least two million-dollar nickel-thieving rings in recent years. In 2018, some miscreant pilfered $10 million worth of cobalt from a warehouse in the Netherlands. But the black market for the red metal seems to dwarf all others.

No one knows exactly how much copper is stolen every year, but it's definitely many, many millions of dollars' worth—possibly billions. A particularly audacious operation took place in 2021, when a Swiss trader brokered a deal for a Chinese company to buy $40 million worth of copper slabs from a supplier in Turkey. When the containers arrived by ship in China, they turned out to be full of nothing but thousands of tons of rock spray-painted a copper color. Even that score is small-time compared to the nearly $200 million worth of copper and other metals stolen in 2023 from Aurubis, Europe's largest producer.

In the United States, the biggest heists are often inside jobs. In 2013, police in Arizona shut down a ring that had ripped off as much as $80 million worth of copper ingots from a mine run by a company called ASARCO. Mine workers in on the scheme would open the gates for trucks driven by their confederates, who loaded up with raw copper and drove right back out. They then sold the metal to recyclers in California, who blackened it to make it look like scrap and then shipped it to China. Unraveling the plot took nerve: at one point, one of the investigators found a goat's head nailed to his door.

Most American copper thieves, however, are small-time opportunists drawn to a laughably easy score. So much copper is just left out in the open. It doesn't take much skill or daring to tear out wiring in an abandoned building, break open an air conditioner sitting unguarded behind an apartment block, or snatch a manhole cover on a quiet suburban street. Thousands of copper thefts are reported each year. The booty includes fire hydrants, a three-thousand-ton bell, a bust of Orville Wright, a statue of Jackie Robinson, and at least one urn containing human ashes.

The damage caused by these robberies often costs far more than the value of the stolen copper itself. Ripped-out copper cables have shut down drinking-water supplies in California, streetlights in Missouri, airport-runway lights in Washington, and whole subway lines in New York City. The US Department of Energy has estimated that copper theft causes $1 billion worth of damage annually to facilities and businesses considered critical infrastructure. Other countries are similarly afflicted. The telecommunications company Telefónica spends more than $18 million every year replacing stolen cable in Argentina. In Spain, police recently arrested sixteen people for stealing fifty-eight tons of copper from a wastewater-treatment plant.

Then there's the shocking number of lives lost. No one knows the exact numbers, but, just from scanning through ten years or so of local news articles, I found dozens of reports of Americans who were fatally electrocuted while trying to steal live copper wire and at least one security guard who was murdered trying to stop one of those thefts.

In South Africa, though, widespread poverty, ineffective police, and soaring metal prices have turned copper theft into a major industry. Criminals siphon off product at every stage of the supply chain. Mines themselves are rich targets. South Africa has huge deposits of platinum, gold, and other precious metals and lots of mines to exploit them. Their deep networks of subterranean shafts and tunnels need power to run lights and digging equipment. That power is carried by miles of electric cable, conveniently left unguarded and out of sight. On any given day, hundreds of desperate people are risking their lives to get that metal.

The thieves are known as zama zamas—roughly meaning "take a chance" in Zulu. These illegal miners clamber down mine shafts, using ropes or handmade ladders, and then make their way into the labyrinth of tunnels. There, they set up underground camps. Hundreds of zama zamas may be living underground at any given time, some spending weeks or even months down in the tunnels. "Down below, temperatures can exceed a hundred degrees, with suffocating humidity. Rockfalls are common, and rescuers have encountered bodies crushed by boulders the size of cars," writes the journalist Kimon de Greef in *The New Yorker*. The men come back to the surface "gray for lack of sunlight," often sick with tuberculosis from inhaling tunnel dust.

Some dig gold and other precious metals out of the rock with hand tools. Others target the copper, stripping it out of cables and other equipment. The zama zamas stash the loot in unused tunnels, hauling it up to the surface at night to transfer it into prearranged vehicles. It's an astonishingly common and deeply disruptive crime. A single mining company, Implats, reported more than eight hundred incidents of cable theft in 2021. Stolen cables have forced companies to shut down mines for weeks at a time.

It's also a phenomenally dangerous way to earn a living. Illegal miners have died by the dozens in floods caused by heavy rains, gas explosions, and other accidents. In 2021, a mining company sealed off a ventilation shaft that a group of zama zamas was using to ferry food and water to their compatriots underground. Desperate, the miners blew open the hole with explosives. Police and private security guards wound up in a pitched battle with the escaping zama zamas. At least eight people were killed.

Above ground, gangs have hijacked dozens of trucks carrying copper to South Africa's ports, making off with millions of dollars' worth of metal. Meanwhile, they are plundering the electric grid so often and so thoroughly that the whole country has been affected. In 2021, the railway company Transnet reported that more than a thousand kilometers of overhead power cables had been stolen. Sprawling over thousands of miles, most of it unprotected, the company's network is an easy target. Even adding guards doesn't always help. A recent report from the Global Initiative Against Transnational Organized Crime notes that "while two security guards may have proved a deterrent in the past," gangs "now come in groups of 20 or 30 and are often heavily armed, with 'spotters' shooting at patrol vehicles. Gangs also dig trenches and throw spikes on the road to hinder responses."

Cell phone towers, water pipelines, and electric-power stations are similarly under assault. Thieves disguise themselves as workers dispatched to tear up underground cables, collaborate with actual power company employees, or just show up with guns and use four-wheel-drive trucks to rip cables out of the ground. In one six-week span in Johannesburg in early 2022, an average of five cable-theft incidents were reported every *day*. All told, the South African government estimates that metal theft costs the nation's economy more than $2 billion each year.

Ordinary South Africans pay a heavy price. Trains are canceled. Power, water, and phone services are shut down for hours or even days at a stretch. A Johannesburg hospital was closed because someone stole its copper pipes, cables, and electrical equipment.

The thievery has lethal consequences, too. Children have died falling into manholes after their covers were stolen. Many of the least cautious larcenists die of electrocution—over two dozen in the last couple of years in Johannesburg alone. Others have been killed by police, or even by other thieves. Police believe a rivalry between gangs involved in stealing cables spurred two mass murders that left a total of twenty-one people dead in the Johannesburg area in early 2022. A number of security guards trying to protect some company's copper have also been wounded or killed—like Moqadi Mokoena, the Johannesburg guard shot to death in his truck.

Fed up with relentless blackouts and water shutdowns, some South Africans have taken matters into their own hands. The wave of copper theft has sparked a backlash of vigilante violence in some impoverished townships. Suspected thieves have been assaulted, beaten, and occasionally lynched. "This is the only language that criminals understand," a resident of a town where a cable thief was beaten to death told local media. But criminals aren't the only victims. In March of 2023, four electric-company workers were killed in a Johannesburg suburb by a mob that mistook them for cable thieves.

Anger over cut-off electricity sometimes feeds into a growing anti-immigrant sentiment, too. Groups like the nativist Operation Dudula movement promote the widespread belief that the thieves are immigrants from neighboring countries. The results are as predictable as they are tragic. In April of 2022, residents of Pimville, a settlement on the outskirts of Johannesburg, were furious over electric-cable theft that had left them without power for days. They blamed immigrants from Lesotho living in a nearby squatter settlement. Operation Dudula activists helped lead a protest march to the settlement, where the marchers were confronted by a group of its residents. Guns were drawn, shots fired. Five people were injured and one killed.

South African authorities and mining companies are flailing for solutions. Railway and power companies are replacing some copper wire with aluminum or copper-manganese alloy wires, but nothing matches copper's conductivity and resistance to corrosion. In a desperate move to discourage theft, the South African government slapped a ban on all copper exports in late 2022, hoping to shut off access to the overseas

markets into which most stolen copper is sold. Judging from the headlines in South African media, though, the thieves barely seem to have
noticed.

And while copper is funding violent criminal gangs in Africa, the
trade in another metal is helping to fund a full-scale war in Europe.

CHAPTER 5

Holding Power

On the morning of February 24, 2022, Russian troops and tanks surged over the border into Ukraine. The world was shocked. Despite months of Russian president Vladimir Putin's threats against the neighboring country, hardly anyone had expected him to launch a full-scale invasion. Ukrainian soldiers fought back ferociously. Rockets and artillery shells filled the air. Highways became battlegrounds. Civilians fled for their lives.

Meanwhile, thousands of miles from the battlefield, a different kind of fear surged through the ranks of the world's commodity traders. The threat they faced wasn't military but economic: The war in Ukraine, they feared, would cut off the tremendous flow of natural resources from Russia to the world's markets. Most traders focused on the country's oil and natural gas, a crucial supply for much of Europe. A few, however, had their gaze fixed on a lower profile but, nonetheless, tremendously important resource: nickel.

Russia is the world's number-one exporter of high-quality nickel. Fearing a supply breakdown and resulting shortages after Russia's attack, commodities traders began snapping up the metal. The price of nickel shot skyward, more than doubling in just twenty-four hours to a record $100,000 per ton. Desperate to calm the frenzy, the London Metal Exchange, the world's premier metal clearinghouse, suspended nickel trading for a week. Billions of dollars' worth of deals were canceled. "It is a very dangerous market right now because this is a market that is not driven by supply and demand, it is driven by fear," Ole Hansen, an executive at the Danish investment bank Saxo Bank, told CNBC at the time.

Why such an uproar over this relatively obscure metal? Because nickel is a crucial ingredient in batteries, the keystone technology of the Electro-Digital Age.

Batteries hold power, in every sense. They are the means by which energy is stored inside electrically powered machines, large and small, ready to be released with the flip of an on switch. Every electric car, truck, bike, and scooter on the roads contains a battery. So does every laptop, mobile phone, digital camera, Bluetooth speaker, drone, and cordless vacuum. It's all but impossible to live a modern life without using something powered by a battery. Today, electric cars are mostly found in developed countries, but batteries are everywhere. An estimated five billion people around the globe own mobile devices powered by lithium-ion batteries.

The billions of batteries in circulation today are just a fraction of the number we're going to have to manufacture in the coming years. We'll need batteries to power our digital machines, of course, but we're going to need exponentially more to make the energy transition possible. No matter how many solar and wind farms you build, if the sun isn't shining and the wind isn't blowing, you have no power. The only way to guarantee on-demand access to energy derived from those sources is to store it in batteries. That goes not only for things that move around, like cars or electric screwdrivers. Homes, factories, offices, and the electric grid itself will also need batteries—big ones. That's why companies like Tesla are already offering building-scale battery systems to homeowners and businesses.

Today, most electric vehicles and digital electronics use some form of lithium-ion batteries. Such batteries comprise three main parts. At one end of the battery is a cathode, typically made of some combination of nickel, cobalt, and manganese. At the other is an anode, usually made of graphite, a form of carbon. (Most of the world's graphite is, unsurprisingly, mined and processed in China. Not all, though. Some comes from a war-torn corner of Mozambique, bedeviled by an Islamist militia linked to ISIS.) Lithium is stored in both the cathode and the anode. When you turn a battery-powered device on, lithium ions flow from the anode to the cathode. In doing so, they generate an electric current that flows out into the device, powering it. When you plug in the device to recharge the battery, the same process happens in reverse, sending the lithium ions back from cathode to anode.

Researchers first began noodling with the concept of a lithium-based energy-storage system in 1912, but the modern lithium battery wasn't invented until the 1970s. Ironically, one of the world's leading fossil-fuel companies was a key funder of the research. At the time, Exxon wasn't worried about climate change. Rather, like many Western corporations (and governments), it was rattled by how the 1973 OPEC oil embargo had exposed its vulnerability to the whims of foreign energy suppliers. So the oil giant recruited a team of scientists to work on alternatives. One of them was a young Stanford researcher named Stanley Whittingham. "We came up with an idea for batteries," he said at a press conference many years later. "I had to go to New York and present it to a committee of the board. They said, 'Sounds good to us!' and within a week they were funding a major project."

At the time, lead-acid batteries—the same kind still used in conventional cars—dominated the market. They were cheap but bulky and heavy. Whittingham and his team came up with a design that used lithium, the lightest of all metals. It allowed them to make batteries that were more energy-dense, meaning they packed the same amount of juice into a much smaller, lighter package. Exxon eventually shut down its battery project, but Whittingham and others carried on the work at other institutions. The concept spread. In the 1980s, as portable electronics became increasingly hot items, manufacturers jumped on the idea of lithium-ion batteries.

Sony brought them to the mass market in 1991 in its consumer camcorders. The compact, powerful batteries meant that video cameras, laptops, phones, and other gadgets could be built smaller and run longer. The batteries quickly ignited a thundering boom in consumer electronics that has yet to subside. Lithium-ion batteries have grown steadily cheaper, more powerful, and more ubiquitous ever since. Whittingham and two other key battery researchers were honored in 2019 with a Nobel Prize. The Nobel Academy declared that the batteries "laid the foundation of a wireless, fossil fuel–free society."

Using these batteries to power cars is a new wrinkle on an old concept. In the early 1900s, in fact, there were more electric cars, mostly running on lead-acid batteries, than their petroleum-powered cousins. But those early batteries weren't great, and fossil fuels were cheap, and so the internal combustion engine took over for the next century.

That began to change in 2008, when Tesla introduced its Roadster, the first mass-produced electric car using a lithium-ion battery. Energy-dense lithium-ion batteries are now reclaiming the ground earlier electric cars lost. Today, gasoline and diesel vehicles still rule the roads, but electrics are catching up faster than almost anyone anticipated. "Gigafactories"— lithium-ion battery–manufacturing plants cranking out enough product each year to store billions of watt-hours of electricity—are springing up all over the world. According to Benchmark Mineral Intelligence, a consulting firm, more than $300 billion in new investments in lithium-ion battery plants were announced between 2019 and 2022.

The United States currently produces less than 10 percent of all lithium-ion batteries, but General Motors, Toyota, Ford, and other corporate leviathans have announced plans to build battery factories across North America in the coming years, supported by billions of dollars' worth of US federal grants and incentives. Boosters talk up an emerging "battery belt" full of electric vehicle–related manufacturing and recycling operations stretching from Michigan down to Georgia and the Carolinas. The new industry is changing the landscape. "The battery plant just north of Commerce [Georgia] is hard to miss," declared a 2023 *Washington Post* article about a newly opened $2.6 billion factory near a small Georgia town. "It looms over Interstate 85 like a monolith: sheer gray walls many stories high, a vast parking lot that extends almost half a mile." Taken together, BloombergNEF estimates, North American battery production will grow sevenfold by 2030.

That will still leave it far behind China. Beijing has been goosing its domestic battery industry for many years, and it has left all others deep in the dust. Electric vehicles are a top priority for China's leadership. As energy historian Daniel Yergin outlines in *The New Map: Energy, Climate, and the Clash of Nations*, electric cars serve at least three purposes for China. One, they help to reduce China's often-overwhelming air pollution. Two, they reduce its dependence on foreign fossil fuels. Three, they offer a way to take the lead in an important economic sector. "As a latecomer to auto production, China would have a hard time catching up and becoming a major competitor in the global market for conventional cars," writes Yergin. "But the electric car is a new game, and there are no big established EV players. Rapid growth of the national EV industry would deliver not only jobs domestically but also the platform to become a formidable exporter and a major force in the global auto industry."

China has done an astonishing job of building up not only its electric vehicle industry, but the whole supply chain for the batteries those vehicles run on. China refines most of the world's cobalt, nickel, manganese, and graphite. It produces more than three quarters of all cathodes and anodes. And it manufactures around 70 percent of all the world's lithium-ion batteries. The world's single biggest battery-making company, Fujian-based Contemporary Amperex Technology (CATL), manufactures nearly one third of all the world's electric vehicle batteries, selling to the likes of Tesla and BMW. Its market valuation is more than those of General Motors and Ford combined. Its spectacular rise has created more billionaires than Google or Facebook. To cap it off, Chinese consumers, encouraged by lavish government subsidies, also buy more electric vehicles than the rest of the world put together.

But one thing China can't control is geology. The refineries and factories and car showrooms may be concentrated within their borders, but the raw metals that make EVs possible are not. Manufacturing all those batteries requires titanic quantities of resources. In fact, electric vehicles and their batteries are the number-one drivers of the growth in demand for critical metals. Beijing has to turn to other countries for many of those resources. High on the list is nickel.

NICKEL

The history of Norilsk, Russia, is nasty, brutish, and short. In the 1920s, explorers discovered deposits of nickel, copper, cobalt, and other minerals on a remote peninsula in northernmost Siberia. Soviet dictator Joseph Stalin wanted them, but finding workers to build an industrial complex and adjoining city in a place where the temperatures are well below freezing for most of the year posed a problem. Stalin solved it by simply forcing prisoners to do the job. An estimated three hundred thousand prisoners cycled through what was dubbed the Norilsk Correctional Labor Camp before it was finally closed in 1955. At least sixteen thousand of them died, taken by malnutrition, disease, and the occasional execution. It was probably little consolation to their families that Norilsk Nickel, the enterprise they helped build, had, by then, become one of the Soviet Union's most important producers of metals, including the one it is named after.

Humans have been using nickel for ages, but we didn't actually realize it until fairly recently. The metal often comes mixed in ore deposits with copper and iron, adding a silvery shine that ancient blacksmiths liked for their tools, weapons, and armor. It wasn't until 1750 that met-allurgists in Saxony came across some copper ore that was lighter in color than usual. They processed it down into an especially bright, sil-very, extremely hard material they hadn't seen before. Flummoxed, they dubbed it kupfernickel, meaning something like "goblin's copper" or "copper with the Devil in it." The following year, a Swedish mineralogist isolated the element and named it nickel. It turned out to be useful for adding luster, reducing weight, and increasing corrosion resistance in tools and weapons. Many countries used it to make shiny, long-lasting coins, including the eponymous US five-cent piece (though that coin is, oddly, made mostly of copper).

Nickel really established its place in the modern world in the 1820s, when scientists figured out that blending it with steel yielded a much stronger, more rust-resistant alloy. Soon, nickel-enhanced steel was being used in guns, ammunition, and vehicles, including to armor war-ships. Then, in 1913, a British metallurgist came up with a mix of carbon, chromium, nickel, and steel that formed a new, rust-resistant compound that was dubbed "stainless" steel. It turned out to be tremendously handy for everything from kitchen faucets to refrigerators to cutlery.

Stainless steel consumes the lion's share of the world's nickel output, but batteries are gaining fast. That's not only because manufacturers around the world are producing more and more batteries containing nickel, but also because many of those manufacturers are increasing the amount of nickel used in those batteries. The more nickel a battery con-tains, the more energy it can store, meaning a car can travel farther on a single charge. The battery in a typical Tesla, for instance, is as much as 80 percent nickel by weight. The battery industry's consumption of nickel jumped 73 percent in 2021 alone. The single biggest source of all that nickel was Norilsk.

After Stalin's death, the new Soviet regime phased out forced labor at Norilsk in favor of salaried workers. When the Soviet Union collapsed and the state's assets went up for grabs, a former Ministry of Foreign

Trade apparatchik named Vladimir Potanin gained control of Norilsk Nickel. The company has since made him into one of Russia's richest oligarchs, boasting a net worth estimated at more than $30 billion. Norilsk Nickel, sometimes known as Nornickel, is today a major producer of cobalt, copper, and platinum, and the world's leading supplier of high-grade nickel. Its customers include battery makers around the world. "The Company produces metals essential for the development of a low-carbon economy and green transport," declared its 2022 annual report—an unfortunately accurate statement.

Along the way, Nornickel earned another distinction: it created one of the most ecologically ravaged places on Earth. "The company's pollution has carved a barren landscape of dead and dying trees out of the taiga, or boreal forest, one of the world's largest carbon sinks," wrote journalist Marianne Lavelle in a 2021 exposé in *Undark* magazine. At the time, Noril'sk emitted as much sulfur pollution as the entire United States. "Satellite instrument readings show that no other human enterprise—no power plant, no oil field, no other smelter complex—generates as much sulfur dioxide pollution," continued Lavelle. "The only entities on Earth that rival or surpass Norilsk's sulfur emissions are erupting volcanoes . . . In volcanoes, however, emissions die down during times of dormancy, while at Norilsk Nickel, the pollution has been steady or rising for nearly 80 years." (The company claims to have significantly cut sulfur emissions since then.) Just to top all that off, in 2020, one of the company's massive fuel tanks collapsed, sending some twenty thousand tons of foaming red diesel cascading into the Ambarnaya River and other waterways. President Putin declared it a federal emergency. Nornickel was forced to pay a $2 billion fine, the highest in Russia's history, and has pledged to spend billions more on pollution control.

Despite all of that, Nornickel remains a major supplier. Nornickel raked in $3.6 billion from nickel sales in 2021 alone. Its importance was made clear by Western countries' quiet decision to exclude it from sanctions for a full two years after Russia invaded Ukraine. In the first year of the Ukraine war, Potanin himself was hit with international sanctions—his assets were frozen, and he was banned from traveling to some countries. But Nornickel's metals were still flowing freely into the world market at the end of 2022, ten months after the invasion of Ukraine. In fact, European Union and US imports of Russian nickel *increased* that year. It wasn't until spring of 2024 that the US and UK finally banned imports of Russian copper, nickel, and other metals.

Western leaders have been acutely concerned about nickel supplies for years. A 2021 US government report warned of a possible impending "large shortage" of high-grade nickel. "Please mine more nickel," Tesla chief Elon Musk begged mining companies in 2020. "Tesla will give you a giant contract for a long period of time if you mine nickel efficiently and in an environmentally sensitive way." Indigenous people in the Norilsk area have written an open letter to Musk, asking him not to buy from the company until it compensates them for turning their ancestral lands into "a lunar landscape." Musk did not respond. Perhaps that's because he and the rest of the nickel industry are shifting more of their attention from Siberia to the South Pacific and the fast-growing nation of Indonesia.

NIRWANA SELLE WAS AN UNLIKELY TIKTOK STAR. THE TWENTY-ONE-YEAR-OLD Indonesian woman's first job out of vocational school was operating a crane at PT Gunbuster Nickel Industry, a Chinese-owned nickel-smelting plant on the island of Sulawesi. She turned the industrial setting into a backdrop for her videos, filming herself at work inside the crane's control room, wearing a hard hat over her hijab and an impish smile on her face as she moved containers of molten metal around or lip-synched along to pop songs. The videos drew millions of views. In her final post, in December of 2022, she rides a motor scooter down a gravel road outside the plant while a saccharine soundtrack plays.

Four days after recording that video, Selle was up in the control room working the night shift when a coal-dust leak sparked an explosion and fire. She and a coworker, Made Detri Hari Jonathan, were trapped inside the room as flames roared up, according to *Vice News*. In a video someone shot at the scene, you can hear screaming. Selle and Jonathan were incinerated. Nothing was left but their bones.

The $2.7 billion plant had only been operating for a year, and yet Selle and Jonathan were just two in a series of workers who had died there. A bulldozer operator had been swept out to sea by an avalanche. Another worker fell into a cauldron of molten slag. The deaths of Selle and Jonathan were the last straw. Hundreds of their Indonesian colleagues went on strike to demand justice and better safety protections. In January of 2023, the strike descended into chaos. The Indonesian workers set fire to dormitories and brawled with Chinese guards. Two workers were killed in the melee. More than five hundred police and soldiers were sent in to restore order. Just six months later, yet another fire at the plant killed one worker and injured six more. Six months after *that*, an explosion at another nearby nickel processing plant, also run by Gunbuster, left eighteen workers dead.

The carnage at the Gunbuster facilities is just one of the horrific side effects of Indonesia's bid to use its vast mineral wealth to boost itself into a top position in the electric-vehicle industry. The nation of nearly three hundred million holds enormous nickel reserves, and for years has been the world's biggest producer of the raw metal. In 2020, however, the government banned exports of unprocessed nickel; only refined products were allowed to leave the country. The goal was to force foreign buyers to invest in building up higher-value refining and manufacturing in Indonesia itself. "We don't just want to build batteries. This is just half of it. We want to build electric cars in Indonesia," President Joko Widodo told *Bloomberg Businessweek*.

From an economic standpoint, the gambit has paid off. Investment has poured in, and production has shot up. As usual, Chinese companies, including the parent company of Gunbuster, lead the pack; according to *Bloomberg*, they have collectively sunk more than $14 billion into nickel refineries, smelters, and other operations on Sulawesi and another nickel-rich Indonesian island. Ford and other non-Chinese multinationals are investing billions more. Tesla has signed multibillion-dollar deals with Sulawesi-based companies and may set up its own manufacturing facility in Indonesia. In the two years following the raw-metal export ban, the value of Indonesia's nickel exports leapt tenfold to $30 billion.

The collateral damage to the environment, however, is prodigious. To clear land for new mines and related infrastructure, more than twenty-one thousand acres of rainforest have been wiped out in one area of Sulawesi alone. That's an area slightly larger than all of St. Thomas, the biggest of the US Virgin Islands. Tailings and other mine waste have polluted streams and waterways. Among the pollutants the mines emit is hexavalent chromium—the same cancer-causing toxin that Erin Brockovich famously campaigned against—which has seeped into drinking water in some areas, according to *The Guardian*. That toxin has also turned up near mines in the Philippines, another major nickel producer. In 2017, more than a dozen nickel mines there were shut down because of the damage they were doing to the environment. Nonetheless, nickel mining is expanding rapidly in the Philippines and has sparked violent clashes between local protesters and police.

On top of the damage caused by nickel mining, that higher-value nickel processing generates even more pollution. Nickel comes in two basic forms: sulfides and laterites. Both require processing to be refined into battery-grade metal, but nickel sulfides require much less processing than laterites. Russia has lots of sulfides. Most of Southeast Asia's nickel, however, is lower-quality laterites. Refining this kind of nickel is considerably dirtier and more complicated. It often involves a process called high-pressure acid leaching, in which the ore is crushed, heated to more than 400 degrees Fahrenheit, mixed with sulfuric acid, and pressurized to separate out the nickel. This method produces millions of tons of acidic waste that need to be handled somehow. Indonesian villagers say runoff from acid-leaching refineries has turned their rivers dark red. One Chinese-owned nickel refinery in Papua New Guinea leaked thirty-five thousand tons of waste into the sea along the country's northeastern coast in 2019, contaminating the water so badly that the government temporarily banned fishing, one of the main sources of local employment. That accidental leak came on top of all the mine tailings that same refinery deliberately and legally dumped directly into the Pacific.

Then there is the filth released into the air. The refineries' smokestacks spew out sulfur dioxide and other pollutants. *The New York Times* reported in 2023 that people living near a Sulawesi nickel plant "complain about the dust pouring off piles of waste, the belching smokestacks, and trucks rumbling past at all hours bearing fresh ore. On the worst days, residents don masks and struggle to breathe." Nickel processing also devours huge amounts of energy, and most of Indonesia's electricity is generated by coal-fired plants. That's right: huge amounts of carbon-intensive coal are being burned to make carbon-neutral batteries.

Despite all of this, other countries in the region are also scaling up their nickel industries. Fully one fourth of all the world's nickel may be in the tiny French territory of New Caledonia, in the South Pacific. That mineral wealth hasn't exactly been a blessing. New Caledonia has been wracked by periodic spasms of violence as the Indigenous Kanak people have battled with descendants of European settlers and other newcomers over the division of the metal booty. Unrest over nickel brought down the local government in 2021.

Meanwhile, the United States is home to only a single nickel mine, located in Michigan. It is expected to be tapped out by around 2026. There's another in the works in Minnesota, but it has been stalled by lawsuits and permitting issues since 2005. In any case, any nickel that is dug out of American soil still has to be sent abroad to be processed. While one or two companies are working to start up domestic nickel refineries, the US currently has none.

Despite all the damage nickel causes, the metal's prospects are good. In fact, several companies are developing batteries that use even more nickel than earlier generations. That's because corporations desperately want to cut their usage of another key battery metal with an even worse reputation.

COBALT

Of all the critical metals, none depends so overwhelmingly on a single source as cobalt. More than 70 percent of the world's supply of this bluish element comes from the Democratic Republic of the Congo—one of the most chaotic nations on Earth, beset by recurring civil wars, endemic corruption, and crushing poverty.

The worst of the DRC's mines are straight out of a nightmare. "The titanic companies that sell products containing Congolese cobalt are worth trillions, yet the people who dig their cobalt out of the ground eke out a base existence characterized by extreme poverty and immense suffering," writes researcher Siddharth Kara in *Cobalt Red: How the Blood of the Congo Powers Our Lives*. "They exist at the edge of human life in an environment that is treated like a toxic dumping ground by foreign mining companies."

The most appalling conditions are in the relatively small-scale, unlicensed operations known as artisanal mines. As many as two hundred thousand people are believed to work in these mines, producing about 15 to 20 percent of the country's total cobalt output. Men wearing nothing but shorts, T-shirts, and flip-flops, with flashlights strapped to their heads, spend their days in deep, cramped underground tunnels, chipping out cobalt with hand tools. Helmets and safety goggles are rare accessories. The dangerous dust they breathe in is the least of their worries.

Not infrequently, the crudely dug tunnels collapse, burying men alive. A single landslide in the late 2010s killed dozens. The miners' harvest is hauled up by ropes, or in sacks carried on their backs, to the surface. There, the ore is sorted, washed, and crushed by women and a shocking number of children.

No one knows exactly how many children work in the DRC's cobalt mines, but the number is definitely in the thousands, perhaps tens of thousands. "Children are required to routinely carry sacks of ore that weigh more than they do," wrote a team of European researchers in a 2019 paper after visiting dozens of DRC cobalt mines. "Children are also often exposed to physical abuse and beatings, whippings, and attempted drownings from security guards, as well as drug abuse, violence, and sexual exploitation." Some are as young as seven years old. Even kids who don't work in the mines are affected by cobalt. Studies have found high levels of birth defects and concentrations of cobalt in the blood and urine of children who live near the mines.

The bulk of the DRC's cobalt comes not from artisanal mines but from large-scale industrial operations, many of which also produce copper. (Copper ore often contains cobalt and nickel, which can be separated out as byproducts.) These operations tend to treat their workers better—but only relatively speaking. According to local non-governmental organizations, exploitatively low pay, dangerous working conditions, and physical abuse are rife in these mines. Those living nearby also suffer. In April of 2023, Amnesty International released a report that found "the expansion of industrial-scale cobalt and copper mines in the Democratic Republic of the Congo has led to the forced eviction of entire communities and grievous human rights abuses including sexual assault, arson and beatings."

Lethal accidents are another constant hazard. In 2018, a truck full of acid overturned in the giant Mutanda Mine, killing eighteen people.

Even more so than nickel, cobalt was of little interest to the world until relatively recently. First mined in 1735, it was initially used mainly as a colorant and later in small amounts for specialized tasks like strengthening jet-engine blades. But in the 2010s, carmakers got interested in electric vehicles and realized that, to build the batteries for those vehicles, they needed cobalt—lots of it. Your smartphone probably contains about a quarter ounce of cobalt; electric vehicle batteries can contain upwards of twenty-four pounds. From 1970 to 2009, global cobalt production averaged around thirty-eight thousand tons per year. The following decade, as demand from battery makers intensified, global production quadrupled.

The surge in demand for cobalt came around the same time that the phenomenal mineral wealth of the DRC was, once again, becoming available to the world, attracting the likes of Robert Friedland, the copper mogul. Any Congolese person who was at all familiar with the country's history must have shuddered as foreigners started coming back in strength. Congo's history of involvement with the global economy is bleak beyond words. The region was colonized in the late 1800s by Belgium, whose rapacious King Leopold II declared himself owner of what he dubbed the Congo Free State, which he converted into a giant rubber plantation worked by legions of enslaved Congolese people. Countless people were mutilated and killed by the Belgian regime. Then, as now, the goal was to extract raw materials and convert them into products to sell to wealthy buyers in foreign lands. Congo's rubber went to feed the booming automobile-tire market, much as its copper and cobalt now feed the booming electric-automobile battery market.

In more recent times, the DRC was convulsed by a series of internal and regional conflicts that killed millions and kept most foreign corporations out. But when the fighting finally simmered down in the early 2000s, mining companies came running back.

Waiting to cut deals with them was an Israeli entrepreneur named Dan Gertler, a former diamond dealer who, years earlier, had befriended DRC president Joseph Kabila. As companies lined up for cobalt- and copper-mining licenses, Gertler connected them with the right government officials—in exchange for millions of dollars in fees for himself. He has, writes Henry Sanderson in *Volt Rush*, "made more money for himself and Congolese elites than anyone since perhaps Belgium's King Leopold II." Gertler has been investigated on charges of bribery and corruption but, so far, has not been convicted of any crimes.

One of the companies most closely tied to Gertler is the Swiss-based multinational Glencore. The company is a major producer of metals critical to the energy transition, including cobalt, nickel, and copper. None of that makes it a "clean energy" company; Glencore is also one of the world's top producers of coal. The commodities giant was founded in 1974 by Marc Rich, an unscrupulous opportunist nonpareil. Rich's resume includes doing business with apartheid-era South Africa and Iran during the American hostage crisis. He was indicted by the US government for tax evasion, racketeering, and fraud, but President Bill Clinton pardoned him in 2001. Rich's style lives on at Glencore, which has admitted to paying millions of dollars in bribes in the DRC.

Chinese construction companies, backed by government credit, also grabbed with both hands at the DRC's riches. Chinese interests, including battery makers CATL, are estimated to own more than 80 percent of the DRC's cobalt-producing mines. Managers in Beijing keep an eye on their workers via live video feeds streamed to their laptops. When it comes to artisanally mined cobalt, Chinese middlemen are the main buyers.

China-based companies have also built up the DRC's roads and ports, the better to get the metals to China. Because regardless of who owns the mine where it is dug up, almost all of the DRC's cobalt ore ends up in China. Chinese refineries handle more than half of the world's cobalt and as much as 90 percent of the ore from the DRC. From China, the processed cobalt is sold on to companies like Sony, Panasonic, General Motors, and Apple. A 2021 White House report declared the cobalt industry "one of the most comprehensive ways China has gained a competitive advantage in the critical materials landscape for batteries."

Based as they are in a country without a free press, Chinese companies have the advantage of not having to worry as much about bad publicity. The working conditions of the DRC's artisanal miners, especially the children, have drawn more attention than any of the other human rights issues connected to critical metals. Amnesty International, *The New York Times, The Guardian, Al Jazeera English*, and many other outfits have all produced damning investigations into the DRC's mines in recent years. Child labor isn't confined to the DRC, of course. According to Aaron Perzanowski's *The Right to Repair*, "roughly a third of the world's tin supply comes from informal Indonesian mines that frequently suffer fatal collapses and employ children. At Cerro Rico, a mine in Potosí, Bolivia, children as young as six years old toil in the deepest, narrowest recesses to retrieve tin, silver, and zinc. Dozens have died in a single year." The US government has also accused China of employing children in its factories making polysilicon for solar panels. There is plenty of shame to go around. Nonetheless, everyone in the mining and electronics industries is acutely aware of cobalt mining's terrible public image. The DRC has promised to reform artisanal mining, but, despite some improvements in recent years, the basic issues around child labor and working conditions remain.

Eager to reassure shareholders that they are addressing the problem, many mobile-phone and car companies have pledged to keep cobalt mined by children out of their batteries. That typically involves hiring third-party auditors to monitor conditions in the mines where their raw materials come from. Critics, however, charge that these inspections are ineffectual. "It's very limited," says Raphael Deberdt, a researcher at the University of British Columbia who used to work for one of the top outfits auditing DRC cobalt mines. "You go to the mine for one or two days, maybe walk around the site, and look at some documents. You check to see if they have policies in place. But you don't really see what's happening," he says.

Another way corporations are trying to keep their hands clean is to simply stop buying cobalt from the DRC. German automaker BMW, for instance, says the company has shifted its purchasing to Morocco and Australia. The industry's eagerness to find an alternative to the DRC is encouraging mining in other places, including Idaho. The state used to be home to America's only cobalt mine, a 10,380-acre facility that was shut down in the 1980s. "By then, the surrounding creeks were lifeless; heavy-metal pollution had killed off most of their fish and aquatic insects," wrote Michael Holtz in *The Atlantic* in 2022. The Environmental Protection Agency designated it one of the country's most contaminated sites, and cleanup work has been going on ever since. Nonetheless, the government has granted at least one new cobalt mine permission to start operating in the state, and several more are in line. No doubt, even with modern regulations, cobalt mines would be hard on the Idaho landscape. But maybe that's a reasonable trade-off to reduce American dependence on Congolese cobalt. When it comes to mining, the choice is never between bad and good but only bad and less bad.

In any case, given the DRC's titanic reserves—the country is estimated to hold nearly as much cobalt as the rest of the world combined—it will certainly continue to be an important supplier for many years. Which, despite everything you've just read, is not entirely a bad thing. Ugly though artisanal mining may be, it is often the only option available. For many Congolese people, the choice isn't between mining cobalt or finding another job; it's between mining cobalt or starving. "There is no other work for most people who live here," a Congolese miner told Kara in *Cobalt Red*. "Yet anyone can dig cobalt and earn money." If all of the battery makers in the world stopped buying cobalt from the DRC tomorrow, they might make their customers and shareholders feel better, but they would also put many thousands of desperately poor people out of work. Every decision has a cost. Every opportunity has drawbacks. A better approach might be for companies to help raise environmental and labor standards in the DRC's mines rather than seeking alternatives to them.

For battery makers, there is yet another option: phasing cobalt and nickel out of batteries completely. "The issue we're all addressing right now is getting the cobalt out," said Whittingham, the scientist who helped invent lithium-ion batteries, in an interview after winning the Nobel Prize, citing the abuses in the DRC's mines as motivation. "There's a huge drive on everybody's part to remove as much of the cobalt as we can out of all these materials."

There are proven battery formulas that don't require either cobalt or nickel. In China, a growing number of electric vehicles run on lithium iron phosphate batteries. These are cheaper to produce but not as energy-dense as their nickel-and-cobalt cousins. They have other drawbacks. Most of the world's phosphate rock is in Morocco and the Western Sahara, where a separatist movement has long simmered. Florida is another significant source, but the Center for Biological Diversity warns that phosphate mines there "displace plants and animals and eat up thousands of acres of valuable habitat that are impossible to restore to their natural state." Phosphate is mainly used for fertilizer production around the world, so a surge in demand from battery makers might inflate prices for farmers. At the moment, the iron and phosphates those batteries contain are so cheap that they might not be worth recycling, says Ryan Castilloux, an analyst with Adamas Intelligence. As a result, they might end up junked en masse in garbage dumps. "The best technology on the market today may be the worst for tomorrow," he says. Everything has a cost.

No matter how much or how little nickel or cobalt is in an electric vehicle's battery, even if it's one that uses iron and phosphate, virtually all of them still need yet another critical metal: lithium.

CHAPTER 6

The Endangered Desert

Deep in the Atacama Desert of northern Chile, the sun beats down on hundreds of enormous shallow rectangular ponds. The patchwork of pools, covering more than seventeen square miles, is colored a range of blues, greens, and yellows, like fabric swatches laid out for Godzilla if he were looking to reupholster his couch. Technically, this is a mine—that is, it's an enormous industrial operation extracting tons of metal from underneath the ground. But there are no diesel-spewing drills or clanking conveyor belts, or even miners, anywhere in view. It's quiet where I'm standing, alongside a teal-colored pool somewhere in the middle of the complex. The only sounds are the rumble of distant trucks and the ambient thrum of pumps pulling up mineral-rich brine from under the desert floor. Most of the extraction is being performed silently, even gently, with nothing but sunshine and gravity working to slowly evaporate the water out of the pools, concentrating it down to an oleaginous yellow-green broth dense with lithium.

"The process is totally natural," says Alejandro Bucher as he escorts me around the facility on a pleasantly warm day in March of 2022. Bucher is the media rep for SQM, the bigger of the two companies currently mining lithium in the Atacama. "We do not add anything to the brine. There's no chemicals involved whatsoever. It's very, very clean."

It seems like a fitting method to extract a substance that is so critical to the transition to "green" energy. Lithium is the irreplaceable key ingredient of the batteries that power virtually all digital devices and electric vehicles. It is the lightest metal in existence, ideal for storing power without adding much weight to a device. Today, nearly three quarters of all the lithium produced worldwide goes into batteries. Demand is soaring. The global market for the soft, silvery-white metal grew sevenfold from 2017 to 2022, approaching a total of $50 billion annually. The International Energy Agency predicts the world will need ten times that much by 2050 to sate the appetite of the Electro-Digital Age.

The Atacama is playing a central role in that ramp-up. The region holds the world's biggest known lithium reserves and supplies more than one fourth of global production. But while the Atacama's mines are helping to save the atmosphere, they may also be killing the Atacama itself.

Here, again, is the familiar conundrum: the spread of digital technology and electric vehicles will ultimately benefit *most* people in *most* places, but the heaviest costs of this shift are being paid by only *some* people and *some* places. Lithium extraction is wreaking havoc the world over. Brine operations in Argentina have reportedly contaminated streams that irrigate food crops. Leaks from a Chinese hard-rock lithium mine have several times turned a river so toxic that it killed not only fish but also cows and yaks that drank from it. Illegal miners have driven villagers from their land at gunpoint in Zimbabwe. Ferocious street protests forced the cancellation of a proposed mine in Serbia. Indigenous peoples, ranchers, and environmental activists are fighting an enormous hard-rock operation slated to open in Nevada, arguing that it will desecrate sacred Indigenous lands and ravage the local ecosystem. And in the Atacama, lithium mines suck up enormous amounts of water in what is already one of the driest places on the planet. The result, independent researchers and many of the Atacama's Indigenous people believe, is that the precious underground water that sustains native plants, animals, and people is disappearing. Lagoons that are home to rare flamingos, vegetation that feeds goats, sheep, and llama-like guanacos, and a way of life followed by Atacameño communities for thousands of years may all be in danger.

Lithium may be the oldest metal in the universe. Scientists believe it was created along with hydrogen and helium in the Big Bang. Humans discovered it quite a bit later, in 1817, thanks to the work of yet another industrious Swedish chemist, Johan August Arfwedson. (What is it with those Swedes?) Lithium turned out to be useful for making a handful of niche products, including ceramic glazes, cold-resistant grease, and extra-strength rubber tires. In 1949, pharmacists learned that ingesting a small amount of lithium helped stabilize the moods of people with bipolar disorder. It's still prescribed today for that purpose. But lithium only came to be produced in large quantities in the 1950s, when the US government started stockpiling it to make one of that era's technological marvels: the hydrogen bomb. The material that makes the H-bomb explode is actually a compound of lithium and hydrogen. The same element that powered my Leaf through the streets of Los Angeles can also be used to blast that city into a smoking crater.

For the next thirty-odd years, nearly all of the world's lithium came from a couple of mines in North Carolina, where it was dug out of hard rock. Then came the discovery of the lithium-rich brines of South America. Guillermo Chong, a now-retired Chilean geologist, was part of the research expedition that found lithium under the Atacama in the early 1970s. "Conditions were very hard," he says, sitting in his home in the Chilean coastal city of Antofagasta. There were no roads in the area back then. At night, he and his teammates shivered in their tents as temperatures dropped to well below freezing. The drilling gear they lugged out in four-wheel-drive trucks was barely up to the task. "The crust was hard as the skin of the devil," says Chong. But after drilling dozens of holes over six weeks in the desert, they knew they were on to something. That something turned out to be the most expansive deposit of lithium ever found to that point. "With geology, you often find nothing," says Chong. "To find something like that—" He rolls his eyes heavenward and lets out an expressive "huf!"

The Atacama lithium mines got going in the 1980s. Extracting lithium from brine was much cheaper than blasting it out of rock, and, as a result, Chile soon became the world's top supplier.

Most of the Atacama is utterly sere and barren, all rock or sand or salt flat painted in muted reds and browns. It's so like the surface of Mars that NASA tests its rovers there, so otherworldly that episodes of *The Mandalorian* have been filmed there.

And yet, it sustains life. The Atacameño people have called it home for at least twelve thousand years. The desert is strewn with pictographs, etched into stone by their ancestors, and the ruins of ancient stone fortresses called *pukkaras*. Today, a handful of Atacameño villages dot the foothills to the east of the lithium mines. They are mostly tucked into ravines that channel rain and snowmelt down from the Andes, ribbons of greenery lush with tamarugo trees, tall grasses, and fields of corn, tomatoes, and other crops. The villagers depend on those streams and the underground aquifers they feed.

The mines sit in a huge salt flat—a *salar*, in Spanish—several miles from the nearest village. Beneath that salt flat lies an enormous reservoir of brine—water thick with salts and minerals, including lithium, that has accumulated over thousands of years. Underground, the fresh and salty bodies of water meet in a mixing zone that forms a kind of border between them. The denser, heavier brines push the lighter freshwater toward the surface, where it forms shallow, brackish lagoons that are home to all kinds of tiny life forms, as well as the rare flamingos that eat them.

The mines slurp up water both sweet and salty. They pipe in fresh water for their employees to drink, wash with, and clean equipment. And they pump up boatloads of brine, at a rate of hundreds of gallons per second, into those colorful surface ponds. The brine is left out in the sun to evaporate over the course of many months. It takes more than one hundred thousand gallons of brine to produce a single ton of lithium.

This process permanently shrinks the total amount of water under the desert. *Engineering & Technology*, a UK-based professional magazine, estimated that, between 1985 and 2017, some 114 billion gallons of water were lost to the environment of the Salar de Atacama due to brine evaporation from just one company.

SQM's position is, essentially, that none of this matters. The fresh water it extracts "has no material impact on the water reserved for drinking and agriculture in the neighboring communities," declares an official company report. The vastly greater quantity of brine the company pumps out, it maintains, doesn't affect the freshwater aquifer or the lagoons it supports. The brine itself "has no environmental value," says Corrado Tore, an SQM hydrogeologist. It's far too salty to drink or even irrigate plants. SQM's scientists maintain a network of hydrogeological sensors that track water levels, and they also monitor the health of the lagoons, vegetation, and wildlife via field studies and satellite imagery. Bottom line, according to their analyses: The mines are causing no damage to the Atacama ecosystem.

Not a single one of the many Atacameño people I spoke with believes that. That includes one who has worked for SQM for many years and others who work for SQM-funded organizations. They all told me that the lithium mines are taking too much water and endangering everything that depends on it.

Water is an especially contentious issue in a region where the taps in villages sometimes run dry and water has to be brought in by truck. It doesn't exactly ease Atacameño worries that, over the years, SQM and US-based Albemarle, the other lithium mining outfit in the Atacama, have each accused the other of extracting more brine than it was legally allowed. A 2016 audit by Chile's environmental regulator found that SQM was indeed overdrawing brine and forced it to cut back. Albemarle may be next. The Chilean government filed a lawsuit against the company in April of 2022, accusing it of over-extracting aquifer water. (Many Atacameños also hold a grudge against SQM, in particular, for what they say was the company's racist treatment of the few Indigenous workers it hired when it began operations in the 1990s. It doesn't help that a son-in-law of Chile's former dictator, Augusto Pinochet, used to be SQM's chairman and still owns billions of dollars' worth of company stock.)

All that said, it's also true that the mining companies provide some jobs in the region, make cash payments to local Atacama communities, and fund development projects, like a small solar-power facility and a water-purification plant. I ask Manuel Salvatierra, the president of the Atacameño council, whether, on balance, considering the benefits they have brought, the mines have been good or bad for his people.

"They've been *really* bad," he says. Thanks to the mines, "there's less forage for our herds. Our agriculture is dying because there's not enough water for our crops." We are sitting at a plastic table in the courtyard of a small local hotel. Salvatierra is dressed in a long-sleeved polo shirt, his black hair tumbling to his shoulders. Splayed in front of him are his cell phone, an iPad, and three USB sticks on a keychain. "I have a cell phone. I use lithium, too. We understand that it may be the solution," he says. "But the way they are getting it is not right. We are being made a sacrifice zone."

Salvatierra's predecessor as president of the Atacameño council is Sergio Cubillos, a slim, youthful activist and long-standing thorn in the side of the mining companies. He has helped organize protests, petitions, a blockade, and at one point went on a six-day hunger strike to protest the mining. He lives in Peine, the village closest to the mines. Most of its few hundred residents live in small homes of cinderblock or cut stone. Chickens and lazy dogs amble around the streets. The locals have mostly traded in the donkeys their parents kept for dusty pickup trucks, but many still graze goats and llamas in the scrub brush down at the desert's edge.

Soon after my visit to the lithium mine, Cubillos takes me, my translator, and a geologist employed by the Atacameño council out to see some of the desert's little-known treasures. We set off from Peine in a four-wheel-drive truck and soon leave the paved road to begin a long, bone-rattling drive over miles of sunbaked brown-and-white salt crust unbroken by even a tuft of grass. Cubillos has lived his whole life in the area, but the internet and his activism have made him a man of the world. As we bounce through the desert, he chatters merrily about the shortcomings of various British soccer teams and the charms of a Canadian mountain resort where he had attended an environmental conference.

Cubillos is concerned about water, not only for its own sake but for the cultural practices it supports. He says his grandparents' generation used to gather flamingo eggs to eat and trade with other tribes, but they had to stop because there were fewer flamingos around. "There was a whole ceremony, to the land, to the water, to the flamingos themselves, to give thanks for the food," he says. "But since the lithium companies arrived, this ancestral practice can no longer be carried out." Brine extraction, he believes, has shrunk the lagoons.

After a couple of hours of following Cubillos's directions to veer left here, and right there, we come upon a little hut made of cactus wood and dried corn stalks. Next to it is an exquisite pool of turquoise water, inlaid in the desert hardpan like a jewel. The vega is maybe twenty feet in diameter, equally deep, and encircled with tall, green grass. Schools of tiny fish swim in its crystal-clear depths. Brilliant-blue dragonflies flit over its surface. A pair of hawks, displeased by our appearance, fly up out of the grass. A few hundred yards away, a couple of wild donkeys stand watching us. "We come here in the summer. We swim. We eat. We spend all day," says Cubillos. "The old people say the mud here is good for your skin." The pool is an utterly unexpected reminder of how much life the desert nurtures and how hard it is to know how much water it holds.

Despite how firmly many Atacameños and their supporters believe that lithium mining is drying out the desert, it's very difficult to assess whether, and to what extent, it actually is. Rainfall and snowmelt fluctuate naturally from year to year, and so do the lagoons and vegetation. The lagoons expand in the winter, when it's cooler and wetter, and shrink in summer. Meanwhile, climate change is also making the region hotter and drier. That makes it tough to get a fix on the overall trend over time and what role the mines play in it. "We don't have long-term baseline studies," says Cristina Dorador, a Chilean microbiologist who has been studying life forms in the Atacama's salt flats for the past two decades. "That makes it complicated."

The region is also cratered with some of the planet's biggest copper mines, including Chuquicamata, which suck up even more water than the lithium mines. In 2021, the Chilean government ordered BHP, owners of the titanic Escondida copper mine, to pay $93 million to compensate for years of overdrawing surface water in the Atacama. BHP has since switched to pumping in desalinated seawater, "but we'll see the effects continue here for another hundred years," says Felipe Lerzundi, who oversees environmental issues for Peine. "We have the impacts of the copper mines added to the impacts of lithium mines."

Much of the data about water in the Atacama is gathered by the mining companies themselves, so it's no surprise that many people don't trust those findings. There is, however, a growing pile of independent research that suggests that the lithium mines are straining the area's freshwater supplies.

A 2019 study by Arizona State University researchers found that soil moisture and vegetation in the salt flat declined between 1997 and 2017, a period during which the lithium mines quadrupled in size. "The expanding lithium industry may be one of the important environmental stressors to the overall health of the local environment," the researchers concluded. An earlier study by a Chilean government agency found that almost one third of the drought-tolerant algarrobo trees on SQM's property were dying, possibly because their roots could no longer find water. Another government study, published in 2018, stated that more water was leaving the salar through pumping and evaporation than was being replaced by rain and snowfall, causing the levels of some wells to drop.

The Atacameños are using some of the money the companies give them to conduct their own research. In recent years, the council has installed sensors to monitor soil and water conditions, gathered satellite imagery, and hired independent researchers. One of those researchers is Javier Escudero Quispe, the hydrogeologist who accompanied us on our tour. He is from the Aymara Indigenous community, a northern neighbor of the Atacameños, and is part of a team investigating whether brine extraction is affecting the lagoons. An amiable thirty-year-old with a round, friendly face, Quispe is from a town several hundred miles to the north but got his degree from a university closer to the Atacama. At one point, he lifts his broad-brimmed hat to reveal a man bun, which he unfurls into a cascade of straight black hair that falls halfway to his waist. "Normally, geologists work for the miners. But that's not my values," he says. When the opportunity came up to work on this project, he signed on eagerly and moved to the desert. "I love the tranquility of this place," he says.

After our visit to the vega, Quispe directs us to one of the lagoons his team is studying. After another hour or so of jolting over the hardpan, we arrive at a place that looks like the surface of another planet.

Glimmering under the desert sun, the lagoon is a vast, irregularly shaped expanse of shallow water as clear as glass. All around its shoreline is a thick crust of what looks like ice but is actually solid salt. There is a faint tang of it in the air. Except for a light wind ruffling the lagoon's surface and the salt crunching under our feet as we walk toward the water, it is completely silent. A flamingo rises up from the lagoon as we approach, flapping off to the far shore to confer with a couple of colleagues.

The only signs of human life are little pipes sticking up out of the water here and there, indicating where Quispe's team had planted sensors to track the lagoon's level and temperature. The team also collects imagery of the lagoons with drones and from satellites. They were still analyzing their data at the time of my visit, but Quispe was confident that their findings would indicate that the lagoons are filled not only with fresh water pushed upward from the mixing zone but also with water from the brine reservoir. (That would echo an analysis by *Engineering & Technology* magazine.) That means, says Quispe, that taking too much brine for the lithium mines could affect the lagoons. A 2022 study by Chilean and American researchers found that surface water, including the lagoons, has declined in the Atacama in recent years.

"This ecosystem is in a very delicate balance. If you disturb it, you can destabilize the whole system," Quispe says. "It could mean less food for the flamingos. With less food, some will die." A 2022 study found the number of flamingos in the Salar de Atacama has declined over the last three decades, though it seems that, for now, they are mostly moving to other lagoons within the wider area that they inhabit.

Bucher says SQM's own research shows that flamingo populations are fluctuating within a natural range and that, overall, the lagoons have not changed in size or depth. Still, pressure on water supplies is likely to increase. Climate change is making the whole region drier, and both SQM and Albemarle plan to expand operations in the coming years. Other companies are also exploring lithium prospects in the Atacama.

"What makes me anxious is if they allow more extraction," says Alejandra Castro, a ranger at Chile's national flamingo reserve who helps run regular censuses of the birds. That, she fears, "could cause an irreversible reduction in the flamingo population."

Thousands of miles to the north, in a different desert in Southern California, Rod Colwell is also extracting lithium from underground brine, but using what he says is a more sustainable method. Colwell heads a startup called Controlled Thermal Resources that is based one hundred miles east of San Diego in the creosote-peppered drylands of Imperial County, near a runoff-filled lake called the Salton Sea. A mile beneath the lake's surface lies a reservoir seething with scalding-hot brine rich with dissolved lithium. (The Atacama's brine is much closer to the surface and not nearly as hot.) The brine's heat has been tapped for geothermal power since the 1980s, but only now are companies starting to extract the white metal.

Colwell is from Australia and has been in Imperial County since 2011, working to get at that underground brine—raising money, schmoozing local politicians, answering questions from regulators and concerned locals. "Our challenge is getting it permitted, getting it built, and getting it out of the ground," he says. "The demand is already there. We could sell two hundred thousand tons a year right now, just to Detroit automakers."

In early 2022, when I met him, he was finally seeing tangible results. From the top of a red-dirt hill about a mile from the sea, Colwell points out the drilling rig that had just come online. It bores down some eight thousand feet to tap into 600-degree-Fahrenheit brine, bringing it to the surface where its heat is converted into geothermal energy and its lithium extracted. The operation was getting closer to full-scale commercial production. Colwell believes it could become a nearly $1 billion business producing at least twenty-five thousand tons of lithium per year.

Rather than letting the water evaporate, as in the Atacama, the company plans to use an innovative chemical process to extract the lithium directly. The leftover salt water will then be pumped back underground. The whole apparatus will be powered with nearly zero-carbon geothermal energy. Taken together, Colwell says, that means his system will take up far less space and use much less water than the Atacama operations. And unlike conventional hard-rock lithium mining—like the controversial mine that is slated to open in Nevada's Thacker Pass—Controlled Thermal's operation won't involve gouging enormous holes in the earth and will generate little in the way of toxic waste and carbon emissions.

It's an attractive pitch. General Motors and Stellantis, the parent company of Chrysler, Fiat, and other automakers, have already signed deals to buy Controlled Thermal's lithium. So has an Italian company that plans to build a $4 billion battery factory in the area. Other companies, including Berkshire Hathaway subsidiary BHE Renewables, are also moving ahead with pilot lithium-extraction projects at geothermal plants by the Salton Sea. All of which has, in recent years, gotten excitable boosters, including California governor Gavin Newsom, to start dreaming out loud about developing the area into a fabulously lucrative "Lithium Valley."

It's a grand vision for such a desolate patch of earth, but America is often fertile ground for those kinds of visions. The whole scenario makes me think of Houston, Texas. At the turn of the twentieth century, Houston was a two-bit town in a backwater state. Then an enormous energy resource—oil—was discovered nearby. Prospectors and drillers flooded in, refineries sprouted up, the oil flowed out, and, over the next several decades, Houston mushroomed into a city of enormous size and wealth (also, pollution and unimaginable sprawl). Pulling an energy-rich liquid out of the ground completely transformed the area and the lives of millions. Why couldn't it happen here?

The chronically impoverished Imperial Valley could certainly use more jobs. Brawley, the largest of the farming towns near the geothermal fields, is a somnolent place where half the storefronts on Main Street are empty or boarded up. What must have once been a charming 1920s-era movie theater is long closed, its paint faded.

Even so, some local leaders are skeptical of the lithium enterprises. "At the end of the day, we all want good businesses, clean businesses, sustainable businesses. But we've been burned many times before," says Luis Olmedo, a longtime community organizer in Brawley. He recalls when solar-power companies moved in promising jobs and prosperity. "They took thousands of acres of farmland that had provided jobs for a hundred years," says Olmedo. "The construction companies came in, built the plant, and were gone," along with those farming jobs.

No one can say whether Controlled Thermal or the other Imperial Valley lithium miners will succeed. Others have tried and failed, and nobody knows what unforeseen effects the industry may have in the long run. What's more, the lithium-production field is increasingly crowded. Australia, already a major mining country, is pushing aggressively to expand its own battery-metals industries, including its vast hard-rock lithium deposits. In 2017, Australia overtook Chile as the world's top lithium producer. Several lithium projects are in the pipeline in the United States, as well, including an effort to reopen the old North Carolina mines. (As usual, no matter where it comes out of the ground, most raw lithium is sent to China for processing.) Mining lithium from rock requires much more energy, however, and therefore generates much more carbon dioxide than extracting it from brine. Hard-rock mining is also typically more expensive.

Even as new sources of lithium come online, given the colossal growth in demand, it's a safe bet that the Atacama's lithium will be an important part of the global supply for many years to come.

The good news in the Atacama is that years of protests, public pressure, and bad press may actually be making a difference. Many companies are now aware that sustainability is an issue they at least need to appear to be addressing: Volkswagen, Mercedes-Benz, and BMW have all signed on to a recently established partnership aimed at supporting "responsible" lithium mining in the Atacama. SQM is also changing its ways to keep up with the times, according to Bucher, making sustainability a top company priority. It's not only the right thing to do, he says, it has become a market imperative; SQM's customers have started demanding it.

SQM has released a sustainability plan that includes pledges to make their operations more transparent, improve relations with nearby communities, and become carbon-neutral by 2040. (The mine currently draws power from a coal-fired plant.) The company participated in a 2023 audit by the Initiative for Responsible Mining Assurance, the US-based nonprofit headed by Aimee Boulanger, the former antimining activist. The group gave SQM's Atacama mine good grades on most measures of worker treatment and environmental responsibility.

Perhaps most importantly, SQM is promising to slash its water use by 2030 (which just happens to be the year when the company's lithium concession comes up for renewal). Bucher says the mine has already cut freshwater consumption in half from pre-2020 levels and will also reduce its brine pumping. It can do that without diminishing production, he says, with efficiency-enhancing techniques such as pumping from areas that have higher lithium concentrations.

Such measures can help, but to Cristina Dorador, the microbiologist, they're not enough. "Lithium mining is not sustainable. It's a myth," she says. "The salars are complex, fragile ecosystems. Every little change can produce big damage." She has long argued that the real problem is much bigger than lithium mining in Chile. The real problem is the sheer amount of natural resources that human beings devour.

"Maintaining the same level of consumption is impossible. It's physically impossible," she says. "We need to change the way that we live on Earth."

One way to look at the Atacama's mines, lithium as well as copper, is to zoom the camera out—up from the ponds and pits, up from the desert, up from Chile—to take in the whole world. From that perspective, it's easier to take a removed and coldly practical perspective: There are very few Atacameños, so, in the grand scheme of things, they simply don't matter as much as the need to fight climate change to protect the whole human race.

Yet, while it is true that the Atacameños' traditional practices linked to the desert's water—growing crops, herding animals, swimming in the vegas—are now practiced by only a small number of people, they are a set of traditions the Atacameños have been following for longer than history holds records. To allow them to disappear is to break a chain that stretches back millennia, to extinguish something that has been kept alive for twelve thousand years. Likewise, the deserts' lagoons and vegas are small. The numbers of creatures that will die if they disappear are relatively few. None of them are of any great importance to the global economy. But they are unique. If they disappear, they will never return.

We've already lost so many ancient traditions and unique natural places to the frenzied development of the modern world. Almost certainly, the Electro-Digital Age will require the sacrifice of still more. The challenge we face is deciding where and upon whom those sacrifices will be inflicted, and figuring out how to minimize those sacrifices as much as possible without slowing down the already overdue energy transition. At the moment, we are asking the Atacameños, Indonesian villagers, Congolese miners, and others like them, who did almost nothing to create the problem of climate change, to bear the heaviest costs to help fix it.

Digging more and more mines in more and more places just won't work as the sole way to meet the need for critical metals. All by itself, this ancient practice can't get us to a sustainable future. It's too expensive, too disruptive, too destructive.

There are other ways, however, to address the question of how to get the critical metals we need. Ways that don't involve the worst harms of land-based mining. Ways that don't involve digging any holes in the ground at all.

The problem is that one of the most prominent of those ways could be even more destructive.

CHAPTER 7

Depth Charge

n early October of 2022, an enormous new creature appeared on the seabed of the Pacific Ocean about fourteen hundred miles southwest of San Diego. It was a remote-controlled, ninety-ton machine the size of a small house, lowered from an industrial ship on a cable nearly three miles long. Once it was settled on the ocean floor, the black, white, and Tonka-truck-yellow contraption began grinding its way forward, its lights lancing through the darkness, steel treads biting into the silt. A battery of water jets mounted on its front end blasted away at the seafloor, stirring up billowing clouds of muck and dislodging hundreds of fist-sized black rocks that lay half-buried in the sediment.

The jets propelled the lumpy stones into an intake at the front of the vehicle, from which they rattled into a steel pipe rising all the way back up to the ship. Air compressors pushed the rocks up, in a column of sediment and seawater, into a shipboard centrifuge that spun away most of the water. Conveyor belts then carried the nodules to a metal ramp that dropped them, with a clatter, into the ship's hold. From a windowless onboard control room, a team of engineers in blue-and-orange coveralls monitored the operation, their faces lit by the polychromatic glow from a hodgepodge of screens.

The ship, called *Hidden Gem*, was a former oil-drilling vessel nearly eight hundred feet long, its deck festooned with ducts, catwalks, cranes, and a helipad. It had been retrofitted for sea mining by the Metals Company, an international firm nominally headquartered in Canada. This was the first full test of its system to collect the ancient black stones.

The rocks are officially known as polymetallic nodules, but the Metals Company's CEO, Gerard Barron, likes to call the nodules "batteries in a rock." That's because they happen to be packed with critical metals, including cobalt, nickel, and copper. Barron's company is at the front of a pack of more than a dozen enterprises slavering over the billions of dollars that could be reaped from those little subsea stones. While mining companies are scouring the Earth's surface for new sources of metals, what might be the richest source of all—the ocean floor—remains completely untapped. The United States Geological Survey estimates that twenty-one billion tons of polymetallic modules lie in a single region of the Pacific containing more of some metals, such as nickel and cobalt, than can be found in all the world's dryland deposits.

"Here's one of them," says Barron when we meet in the lobby of a chic Toronto hotel in the summer of 2022. He casually pulls one of these prehistoric oddities out of his jacket pocket and hands it to me. It's a fist-sized black lump covered with knobby little protrusions, like a sloppily formed hamburger patty. Barron is a muscular Australian in his mid-fifties with swept-back dark hair and craggy, Kurt Russell–esque looks.

Barron had flown in from London for a mining conference. He's been doing this kind of thing for years, traveling the world to talk up deep-sea mining to investors and government officials. He and other would-be sea miners argue that collecting nodules from the deep will not only be cheaper than traditional mining but also gentler on the planet—no rainforests uprooted, no child labor, no toxic tailings poisoning rivers.

Barron has been working for many years to get mega-scale mining going on the ocean floor, and the goal is almost within his grasp. The Metals Company has tens of millions of dollars in the bank and partnerships with major maritime and mining companies. *Hidden Gem*'s test run in 2022 marked the first time since the 1970s that any company had successfully tested a complete system for harvesting nodules. The main thing holding them back, at this point, is international law, which currently forbids deep-ocean mining. That, however, may soon change. In 2021, the Metals Company teamed up with the tiny South Pacific island nation of Nauru to trigger an obscure legal process that could let them bypass the international prohibition and get a license to start full-scale operations as soon as late 2025.

That prospect has sparked an outraged backlash. Environmental groups, scientists, and even some corporations in the market for battery metals fear the potential havoc that could be unleashed by digging up the seafloor. The oceans provide much of the world's biodiversity and nearly one fifth of all the animal protein that humans consume each year and are the planet's most important repository of carbon. No one knows how such an unprecedented incursion would affect the many life-forms that live in the abyssal depths, the marine life farther up the water column, or the ocean itself.

The European Parliament and countries including Germany, Chile, Spain, and several Pacific island nations have joined dozens of organizations in calling for at least a temporary moratorium on deep-sea mining. Several banks have declared they won't lend to ocean-mining ventures. Corporations including BMW, Microsoft, Google, Volvo, and Volkswagen have pledged not to buy deep-sea metals until the environmental impacts are better understood. Even Aquaman is opposed: actor Jason Momoa narrated a 2023 documentary denouncing sea mining.

"This has the potential to transform the oceans, and not for the better," says Diva Amon, a marine scientist who has worked extensively in the area of the Pacific primarily targeted for mining, including as a contractor for one of the sea-mining companies. "We could stand to lose parts of the planet and species that live there before we know, understand, and value them."

None of that deters Barron. The times demand urgent action, he says. "The biggest challenge to our planet is climate change and biodiversity loss. We don't have a spare decade to sit around," he tells me. By the end of *Hidden Gem*'s 2022 trial, the vehicle had collected more than three thousand tons of the stones, mounded up in the ship's hold in a glistening black pyramid nearly four stories high. "This," Barron promised the press, "is just the beginning."

The nodules have been growing, in utter blackness and near-total silence, for millions of years. Each started out as a fragment of something else—a tiny fossil, a scrap of basalt, a shark's tooth—that drifted down to the plain at the very bottom of the ocean. In the leisurely unfolding of geologic time, specks of water-borne nickel, copper, cobalt, and manganese slowly accreted onto them. By now, trillions of them have come into being, half-buried in the sediment that carpets the ocean floor.

One March day in 1873, some of those subaqueous artifacts were dragged into the sunlight for the first time. Sailors aboard the HMS *Challenger*, a former British warship retrofitted into a floating research lab, dredged a net along the sea bottom, hauled it up, and dumped the dripping sediment onto the ship's wooden deck. As the expedition's scientists sifted through the mud and muck, they noted with interest the many "peculiar black oval bodies" contained therein. They soon determined the rocks were concretions of valuable minerals. It was a fascinating discovery, but it would be almost a century before the world began to dream of exploiting the stones.

In 1964, American geologist John L. Mero published an influential book called *The Mineral Resources of the Sea*, which estimated that the nodules contained enough manganese, cobalt, nickel, and other metals to feed the world's industrial needs for thousands of years. Mining the nodules, Mero speculated, "could serve to remove one of the historic causes of war between nations, supplies of raw materials for expanding populations. Of course it might produce the opposite effect also, that of fomenting inane squabbles over who owns which areas of the ocean floor."

The concept struck a chord in an era when population growth and an embryonic environmental movement were fueling concerns about natural resources. Seabed mining was suddenly hot. Throughout the 1970s, governments and private companies rushed to develop ships and rigs to pull up nodules. There was so much hype around the industry that it seemed plausible when, in 1974, billionaire Howard Hughes announced that he was sending a custom-built ship out into the Pacific to search for nodules. (In fact, in a bit of James Bond–esque skullduggery, the CIA had secretly recruited Hughes to provide a cover for the ship's real mission: to covertly retrieve a sunken Soviet submarine.) But none of the actual sea miners managed to come up with a system that could do the job at a price that made financial sense, and the fizz went out of the nascent industry.

By the turn of the twenty-first century, advancing marine technology made sea mining seem plausible again. With GPS and more sophisticated motors, ships could float above precisely chosen points on the seafloor. Remotely operated underwater vehicles grew more capable and dove deeper. The nodules now seemed to be within reach, just as China and other booming economies were ravenous for metals.

Barron saw the potential bonanza decades ago. He grew up on a dairy farm, the youngest of five. "I knew I didn't want to be a dairy farmer, but I loved dairy-farm life," he says. "I loved driving tractors and harvesters." He left home to go to a small regional university and started his first company, a loan refinancing operation, while still a student. He found he was a good entrepreneur. After graduating, he moved to the state capital, Brisbane, "to discover the big wide world." Over the years, he's been involved in magazine publishing, ad software, and importing conventional car batteries from China.

In 2001, a tennis buddy of Barron's—a geologist and former prospector named David Heydon—pitched him on a new company he was spinning up, a sea-mining outfit called Nautilus Minerals. "I was fascinated to learn that the oceans were filled with metals," recalls Barron. "I figured that had to be a better way of getting them than land-based mining." He put some of his own money in to the venture and rounded up other investors.

Nautilus wasn't going after polymetallic nodules but rather what seemed like an easier target: underwater formations called seafloor massive sulfides, which are formed by geothermal vents in relatively shallow waters and are rich in cobalt and other metals. The company struck a deal with the government of Papua New Guinea to mine sulfides off their coast. (Under international law, countries can do basically whatever they want within their own territorial waters, which extend up to two hundred miles from their coastlines.) It sounded good—good enough, in fact, to attract half a billion dollars from investors, including Papua New Guinea itself.

But in 2019, after spending nearly $460 million, the project collapsed in the face of local resistance and financial troubles. Nautilus went bust. Neither Barron nor Heydon lost any of their own money, though. Both had sold their shares about a decade earlier, with Barron clearing about $30 million in profit. Papua New Guinea, where some 85 percent of the population lives in poverty, was out $120 million. "It wasn't my business," Barron says. "I was just supporting David, really."

Heydon, meanwhile, was building a new company called DeepGreen Metals—rebranded in 2021 as the Metals Company—this time pursuing polymetallic nodules. By then, the growing demand for electric vehicles had added both a new potential market and an extra environmental justification for the project. Barron came on as CEO, and several other Nautilus alums also joined, including Heydon's son Robert. Along with other would-be miners, they started knocking on the door of a little-known but extremely important agency called the International Seabed Authority.

It's more than a little disturbing that the world's main instrument for regulating seabed mining is a small group of multinational bureaucrats based in Kingston, Jamaica, tasked with interpreting a few paragraphs in a sprawling, decades-old treaty. The International Seabed Authority has the contradictory task of protecting the ocean floor while organizing its commercial exploitation. Back in the 1980s, most of the world's nations—notably excluding the United States—signed a kind of constitution for the oceans, the United Nations Convention on the Law of the Sea. Among many other things, the document established the International Seabed Authority to represent what are now its 168 member nations. The organization was charged with devising rules to govern the then-nonexistent deep-sea mining industry. It has been working—very, very slowly—to develop those rules ever since. A forest of thorny issues, from environmental protections to profit sharing with poorer countries, has yet to be resolved. Until regulations are agreed upon, full-scale mining is prohibited. But in the meantime, the agency can grant miners the right to explore specific areas and reserve them for commercial exploitation. The ISA also declared that private companies must partner with a member country. Even the tiniest member country will do.

As of 2023, the agency has granted permission to twenty-two companies and governments to start exploring enormous swathes of the Pacific, Atlantic, and Indian Ocean seabeds. Most are targeting polymetallic nodules lying roughly three miles underwater in the Clarion-Clipperton Zone, an expanse of the Pacific between Mexico and Hawaii measuring 1.7 million square miles. Holding the rights to three of the choicest parcels is Gerard Barron and the Metals Company. The company's chief financial officer told investors that those parcels could yield metals worth $31 billion.

Here's what makes all of this urgent. The mining ban has a loophole: the two-year trigger. A section of the treaty known as Paragraph 15 states that if any member country formally notifies the Seabed Authority that it wants to start sea mining in international waters, it starts a clock ticking. The organization will have two years to adopt full regulations. If it fails to do so, Paragraph 15 says, the ISA "shall none the less consider and provisionally approve such plan of work." This text is commonly interpreted to mean mining must be allowed to go ahead, even in the absence of full regulations. "Paragraph 15 was appallingly drafted," says Duncan Currie, a lawyer for the Deep Sea Conservation Coalition, an international umbrella organization of dozens of groups. "Several countries dispute the idea that it means they need to automatically approve a plan of work."

In the summer of 2021, the president of Nauru formally notified the Seabed Authority that his country, along with the Metals Company's wholly owned subsidiary Nauru Ocean Resources, planned to begin sea mining. The two-year trigger had been pulled.

"As an environmentalist myself," says Barron, he finds the opposition from environmental groups frustrating. "'Save the oceans' is a really easy slogan to get behind. I'm behind it!" he says. "I want to save the oceans, but I also want to save the planet." It might be true that getting metals from the seafloor is less damaging than getting them from land. But so far, few outside the industry are convinced.

For starters, very little is truly known about the deep ocean. Gathering data hundreds of miles from land and miles below the water's surface is extraordinarily difficult. A single day's serious research can cost up to $80,000, and sophisticated tools such as remotely operated vehicles have only recently become available to most scientists. In 2022, thirty-one marine researchers published a paper that reviewed hundreds of studies on deep-sea mining and concluded that there are still huge gaps in our knowledge. The authors also interviewed twenty scientists, industry members, and policymakers; almost all said the scientific community needed at least five more years "to make evidence-based recommendations" on how the industry should be regulated.

Every phase of the mining process entails serious risks for the world's oceans, which are already stressed by pollution, overfishing, and climate change.

Start at the bottom. A fleet of machines, each weighing dozens of tons, lumbering on tank treads over the pristine ocean floor and prying loose thousands of nodules from sediment that has lain undisturbed for millennia is inevitably going to cause some damage. According to *Scientific American*, over the course of a standard thirty-year contract, a single one of those machines could strip 3,900 square miles of seabed—an area larger than Rhode Island and Delaware combined.

Corals, sponges, nematodes, and dozens of other organisms live on the nodules themselves or shelter beneath them. Other creatures float in the water around them, such as anemones with eight-foot-long tentacles, rippling squidworms, millennia-old glass sponges, and ghostly white Dumbo octopuses. "It's like Dr. Seuss down there," says Amon, the marine scientist. The nodules, she believes, are a critical part of the ecosystem that supports all those creatures. "Those nodules formed over millions of years," she says. Whatever the fallout of removing them is, "the damage is, in effect, irreversible."

Some scientists are also concerned that huge amounts of carbon embedded on the ocean floor could be released and potentially interfere with the ocean's ability to capture and store more. "I've been down there," deep-sea explorer and investor Victor Vescovo told Bloomberg News. "I've seen the polymetallic fields, and there is no way to extract polymetallic nodules from the seafloor without utterly annihilating the bottom of the seafloor. It cannot be done."

Silt and sand stirred up by the collector vehicles will also rise up, creating plumes of sediment that could cloud the water for many miles, linger for months, and suffocate creatures that live higher up in the ocean. Those plumes might also contain dissolved metals or other toxic substances that could harm aquatic life.

Moving upward, the noise and light emitted by the harvester vehicles and machinery for lifting the modules to the surface could affect any number of creatures who have evolved to live in silence and darkness. A recent study found that the racket from just one seabed-mining operation could echo for hundreds of miles through the water, potentially interfering with the ability of underwater organisms to navigate and find food and mates. "It's not just going to be the seafloor and the animals that live there that will be affected," says Amon. "It's also going to be a lot of life from the sea surface to the seafloor."

Once the nodules have been carried up to a ship, the silt-infused water that accompanied them will have to be dumped back into the sea, creating another potentially dangerous sediment plume. "We are talking about massive volumes. Fifty thousand cubic meters a day," says Jeff Drazen, a University of Hawaii oceanographer who has also worked extensively in the Clarion-Clipperton Zone, including on a research mission funded by the Metals Company. "That's like a freight train of muddy seawater every day."

A 2022 report from the United Nations Environment Programme sums up the grim picture: "Current scientific consensus suggests that deep-sea mining will be highly damaging to ocean ecosystems." More than eight hundred marine-science and -policy experts have signed a petition calling for a "pause" on sea mining until more research has been conducted.

Barron insists that his company is committed to getting the science right. He points out that it has funded more than a dozen research expeditions (which were required by the Seabed Authority). By now, he argues, we know enough. "The lack of full scientific knowledge should not be used as an excuse not to proceed when the known impacts of the alternative—land-based mining—are there for us all to see," he says. Sea mining will be less destructive, he insists. "That's a certainty."

Whoever authored the Metals Company's own registration filing with the Securities and Exchange Commission, however, wasn't quite so sure. That document notes that nodule collection in the Clarion-Clipperton Zone is "certain to disturb wildlife" and "may impact ecosystem function" to an unpredictable extent. The filing adds that it may "not be possible to definitively say" whether nodule collection will do more or less harm to global biodiversity than land-based mining. In any case, it's not an either-or proposition. At least some terrestrial mining will certainly continue for decades, no matter what happens under the sea.

The Metals Company's critics say the company isn't interested in what the science shows. One environmental scientist quit a contract job with the enterprise, complaining, in a since-deleted LinkedIn post, that "the company has minimal respect for science, marine conservation, or society in general . . . Don't let them fool you. Money is the game. It's business in their eyes, not people or the planet." (Barron says this person was a disgruntled former employee and his charges aren't true. My efforts to contact him were unsuccessful.)

The Metals Company is the only deep-sea mining outfit that is not backed by a major corporation or national government. It's a startup, wholly dependent, to this point, on fickle investor capital. That could certainly explain why Barron seems to be in such a hurry to start mining. When I ask him why the company triggered the two-year rule, he interrupts to clarify: "Well, Nauru did. We didn't. Nauru did."

You'd have a hard time finding a more extreme example of despoliation of a tropical paradise, of a Fall from Eden, than Nauru. When the first European ship came across this isolated, eight-square-mile island in the South Pacific in 1798, the captain was so charmed by the fair weather, lovely beaches, and the locals' friendly welcome that he named it Pleasant Island. But once an Australian geologist discovered that the spot was loaded with high-grade phosphates, which were in high demand as fertilizer, the outside world rushed in. Over the course of the twentieth century, the island was strip-mined to the brink of oblivion. Its once-lush interior was reduced to what *The Guardian* described as a "moonscape of jagged limestone pinnacles unfit for agriculture or even building." By the 1990s, the phosphates began running low, and Nauru, a nation of roughly ten thousand people, needed another source of income. The island tried to set itself up as a no-questions-asked offshore-banking haven, but so much suspicious cash poured in from Russian crime syndicates and other unsavory types that international pressure forced the country to tighten its regulations. The island's next moneymaking scheme was to rent some of its territory to Australia to use as an immigrant-detention center. Over the course of several years, detainees protesting conditions there rioted, staged hunger strikes, and sewed their lips shut. One set himself on fire and burned to death. Australia finally moved all the detainees elsewhere in 2023.

Given all that, it's not hard to see the appeal for Nauru of teaming up with the Metals Company, especially since the mining zone is nowhere near the island. "Our people, land and resources were exploited to fuel the industrial revolution elsewhere, and we are now expected to bear the brunt of the destructive consequences of that industrial revolution," including sea-level rise, wrote Margo Deiye, Nauru's representative to the UN, in a 2022 newspaper op-ed explaining why her country is supporting sea mining. "We're not sitting back, waiting for the rich world to fix what they created."

Barron, who has never set foot on the island, insists that the relationship between the Metals Company and Nauru is a respectful partnership, not a modern version of colonial exploitation. "It's horrible what happened to Nauru," he says. "They were absolutely fucked over by the Germans, the English, the Australians, and the Kiwis." He says he's proud of the benefits the Metals Company is bringing to Nauru, including the more than $200,000 the company has doled out to support community programs on Nauru and on Kiribati and Tonga, the two other small Pacific nations with which it has business arrangements. "The real contribution," he adds, "will be when we start paying royalties," the partner nations' yet-to-be-decided percentage of sea-mining revenues.

The Metals Company's own finances, however, are not exactly robust. Barron took the company public on September 10, 2021, a few months after the two-year rule was triggered, claiming he had commitments totaling $300 million from investors. Its stock topped $12 a few days later.

But two key investors never delivered, leaving Barron and his team with only a third of their expected capital. The company's stock price plummeted, and it has hovered around $1 ever since. The company is suing the faithless investors and is being sued itself by other investors who claim they were misled by the company. In May of 2023, Maersk, a shipping industry giant that had invested in the Metals Company for several years, apparently lost confidence in the enterprise and sold off its shares.

Meanwhile, the Metals Company has burned through some $300 million. A substantial chunk of that cash wound up in Barron's pocket. He and his partner, Erika Ilves, whom Barron hired as chief strategy officer, were, together, paid more than $20 million in salary and stock options in 2021.

(Before she moved into sea mining, Ilves worked in the even more exotic field of space mining—as in outer space. The thousands of asteroids orbiting between Mars and Jupiter contain unimaginable troves of metals, and starry-eyed entrepreneurs have been trying to get at them for years. In the mid-2010s, a pair of asteroid-mining startups attracted investment from the likes of Google founder Larry Page and executive Eric Schmidt and filmmaker James Cameron, but both projects soon flamed out. Undeterred, a handful of startups are trying to pick up the torch. "Asteroid mining would have no impact on our biosphere. You can strip-mine them or blow them up—they're just dead, lifeless rocks," says Mitch Hunter-Scullion, the young CEO of the UK's Asteroid Mining Corporation. "Someone's got to get them. Why not me?")

Journalists at *Bloomberg* and some environmental organizations have suggested that the Metals Company holds unfair leverage over its partner nations and have drawn attention to the seemingly cozy ties between the company and the International Seabed Authority—in particular, the ISA's secretary general, Michael Lodge. A 2022 *New York Times* investigation alleged that the ISA gave the Metals Company's executives access to data indicating where the most valuable seabed tracts were located, then helped the company secure the rights to those areas. Both the agency and the company say that all their dealings have been legal and appropriate. (Lodge also made his stance on environmentalists pretty clear, telling the *Times*: "Everybody in Brooklyn can say, 'I don't want to harm the ocean.' But they sure want their Teslas.")

Between Barron's outspokenness and his company's legal and financial pyrotechnics, the Metals Company has drawn most of the media coverage around sea mining. The other players are keeping a lower profile, perhaps deliberately. "TMC is very bold, but the other companies are piggybacking on them," says Jessica Battle, who heads the World Wildlife Fund's campaign against sea mining. "Once one license is given, others will follow." Belgian maritime giant DEME, shipbuilder Keppel Offshore & Marine, and the governments of South Korea, India, Japan, Russia, and China have launched dozens of research expeditions in recent years. China is particularly eager: It has three state-affiliated companies licensed to explore ninety-two thousand square miles of seabed for polymetallic nodules in the Pacific, is actively building its deep-water mining capacity, and opposes the idea of a sea-mining moratorium. Beijing contributes significant funds to the ISA and wields considerable clout there, according to *The Washington Post*.

DEME's sea-mining subsidiary, Global Sea Mineral Resources, is also well positioned to take the lead if the Metals Company stumbles. "They've got the backing of a multibillion-dollar company and access to European resources for design," says Currie, the environmental lawyer. "They can wait ten or fifteen years and it wouldn't be the end of the world for them. Whereas with the Metals Company—look at their stock price. If their license isn't approved, it's hard to see how they survive." Global Sea Mineral Resources has also been running extensive test runs in the Pacific— and learning its own lessons about how badly things can go wrong.

A FRANTIC KNOCKING ON THE METAL DOOR OF HIS CABIN JOLTED KRIS DE BRUYNE awake. It was early in the morning of April 25, 2021, and De Bruyne, a young Belgian engineer with Global Sea Mineral Resources, was aboard an industrial ship far out in the Pacific. De Bruyne was helming a team of researchers testing the Patania II, a bright green robotic collector-vehicle prototype, similar to the machine deployed by the Metals Company. Now, one of his team was shouting through the door to him: "Something really bad happened! The umbilical disconnected!"

That was, indeed, *really* bad. The umbilical is a Kevlar-jacketed cable, stuffed with fiber-optic and copper wire, three miles long and as thick around as a person's arm. It was the only thing tethering the Patania II to the ship.

"Is it going down?" De Bruyne called back.

"Yes!"

De Bruyne scrambled into his red coveralls and ran up on deck. The crew had been hauling up the vehicle after a test dive when, just fifty feet from the surface, the umbilical snapped. The thirty-five-ton vehicle went spiraling down to the bottom of the Pacific.

Luckily, the Patania II landed tracks-down, with its locator system intact and sending acoustic pings up to the ship to indicate where it was. It took a couple of days, but De Bruyne's crew was able to send down a small submersible robot equipped with three-fingered, Doctor Octopus–style tentacles that reattached the repaired umbilical, allowing them to bring the machine back to the surface.

"It was relatively easy. Well, I say it was very easy, but it was also like 'AAAAHHH!!!' and 'NOOOO!'" recounts De Bruyne at DEME's headquarters near Antwerp, Belgium. "It was an emotional roller coaster."

When they hauled the Patania II up, they found it almost completely undamaged. To De Bruyne, the snapped cable was just one of what he calls the "teething problems" that typically come with launching such a complex piece of machinery. Not everyone was so sanguine. Earlier in the expedition, De Bruyne had also had to contend with Greenpeace activists, who had sailed up alongside his ship and painted RISK! on it in huge yellow letters.

De Bruyne has a fanboy's enthusiasm for his job, eager to tell me all the details about different bits and pieces of machinery and how they all work together. He's also acutely conscious of the criticism directed at his industry and seems to take it personally. His parents were traveling veterinarians and raised him and his brother in Rwanda and Vietnam. "I grew up in nature. I'm not the nature destroyer they want me to be," he says. "The non-governmental organizations and the environmentalists, they forget that we also have our stories, and that we want to do something good for the world, as well."

The Patania II mission, he points out, was accompanied by a separate ship full of marine scientists who monitored the machine's effects on the ocean (as was the Metals Company's foray). "Once in a while, I'll ask myself: Am I still doing the right thing?" he continues. "I still think we're doing the right thing, because we're still doing research." He says he's not even convinced deep-sea mining should go ahead. "We need to know what the impact would be of deep-sea mining, and I'm contributing to getting answers to that question. That's how I feel about it."

Global Sea Mineral Resources has already sunk at least $100 million into developing its subsea mining system, and it recently announced a partnership with Transocean, a major offshore oil-drilling outfit. The sea-mining company is now designing the much larger Patania III—the first of what the company hopes will be a fleet of full-scale sea-mining robots that will hit the ocean floor around 2028.

That might give researchers enough time to develop the scientific understanding needed to craft regulations to safely mine the seafloor—or to determine if it should be done at all. Even if all the environmental issues can be addressed, seabed mining might simply become superfluous if, for instance, lithium iron phosphate batteries take over the industry and the world no longer needs as much cobalt and nickel.

Gerard Barron is not planning to wait. "Got the boat, got the machine, announced the partnerships on how we're going to process the nodules," he says. Assuming the Metals Company gets the go-ahead from the Seabed Authority, he says, everything is on track to start harvesting nodules by late 2025. The company's goal for that year is three million tons, scaling up to several times that amount in the next decade.

The two-year deadline expired on July 9, 2023. The Seabed Authority had not, by then, developed a code to govern sea mining. It announced it would continue working on the issue and aim to come up with a code sometime in 2025. Nonetheless, it is now obligated to begin accepting applications for commercial-mining permits. It's still an open question as to whether it must approve any of those applications before the full set of regulations is in place. "No one," says Currie, the ocean conservation lawyer, "is sure how this will play out."

Regardless of what the Seabed Authority decides, however, it's very likely some type of ocean mining will get underway in the near future. Remember that individual countries don't need anyone's permission to dig up metals that lie in their territorial waters. The Norwegian government is considering opening a swathe of the Arctic Ocean bigger than the United Kingdom to sea miners. Japan has already successfully tested a system for extracting rare earths from deep-ocean mud and is moving toward full-scale extraction (with the explicit goal of reducing its dependence on China for those particular metals). The Cook Islands, a South Pacific nation, has authorized Global Sea Minerals Resources and two other mining companies to explore its waters for polymetallic nodules. Even Papua New Guinea, where the Metals Company's predecessor, DeepGreen, went bankrupt, is talking to another mining company about restarting ocean mining there.

Everything has a cost. There is no question that sea mining will cause harm; what's unknown is how serious that harm will be and how far it will spread. On balance, it might—*might*—turn out to be less harmful than land-based mining. But no one can know for sure unless and until it actually happens. And at that point, we risk damaging the ocean as badly and irreparably as we've damaged the atmosphere.

What often gets forgotten or overlooked, however, is that finding new sources of metals, whether on land or in the sea or even in outer space, isn't an end in itself. It's a means to the end of having enough of the metals we need to provide a decent way of life. There are several completely different ways through which we might be able to reach that goal, or at least come closer to it, without risking the oceans or devastating the land.

PART TWO

THE REVERSE SUPPLY CHAIN

A nation is not made wealthy by the childish accumulation of shiny metals, but is enriched by the economic prosperity of its people.

—ADAM SMITH, EIGHTEENTH-CENTURY ECONOMIST

CHAPTER 8

Mining the Concrete Jungle

teve Nelson grabs the lip of the dumpster with his thick, calloused fingers, scrambles nimbly up the side, and drops down into the trash inside. He's fifty-seven, and his many years of living on the streets of Vancouver, Canada, show in his frazzled gray hair, weathered face, and unruly teeth. But he's otherwise in great shape, sinewy, strong, and full of good cheer.

The dumpster sits behind a warehouse in an industrial neighborhood in central Vancouver. It fills up every day with all kinds of random junk. "OK, let's see what we got," says Steve, knee-deep in a gumbo of heavy plastic trash bags, bits of broken machinery, and cast-off metal miscellany. He gives the bags a grope. "Doesn't feel like any wire," he mutters. A quick scrounge through the loose metal pieces, though, turns up a few lengths of electrical wire, some small sheets of aluminum, and a large outdoor light fixture, the kind you might see on top of a pole in a parking lot. That's a good find. "I can take this apart without a tool and make it into money!" Steve crows. He shows me how to remove a small corner piece of metal and use it as an improvised screwdriver to loosen the casing on the light. "Then you just drop it on a corner, and it'll crack at the welds," he explains. "The inside will pop out, and you've got yourself the copper core." Copper is the desideratum of every scavenger like Steve, the most valuable common scrap metal. There's probably two pounds of it inside the light, Steve estimates. There's a similar amount of cast aluminum in the casing. At the price those metals were fetching on this particular chilly May day, Steve figures he'll net about three bucks and change when he sells this haul to a scrapyard, plus a bit more for the aluminum sheets and the copper he'll strip out of the electrical wire. Not bad for a few minutes' work.

He's not going to take his finds apart here, though. That would make a mess, and Steve always makes sure not to do that. The people who work in the warehouse and the other nearby businesses know him, know that he's careful that way, and so they don't hassle him if they find him rummaging through their trash. He even cleans loose screws and scraps off the ground with a magnet he carries. Some of the regulars on what he calls his trapline even save their junk to give to him. "They can make money themselves off the metal, but they don't mind. They've seen me out there in the snow and the rain for twenty fucking years," he says. "They make my life a little easier."

He lugs his finds over to where we left our bicycles. Steve does all his scrapping on a bike, piling the bits and pieces on a jerry-rigged trailer. He has loved riding ever since his days as a bicycle messenger in the 1980s. Steve's been through some tough times since, and his current lifestyle is not one many would choose. But he does still have a bit of outlaw panache to him, a guy making his way outside the system, living by his wits.

It looks to me like Steve's trailer is already packed to capacity, bungee cords straining to hold the mismatched collection of metal detritus. Somehow, he finds space for the new scraps, cramming them in here and there and using an old bike-tire inner tube to tie the light fixture on top of the whole mess. He never uses a shopping cart. "Having a shopping cart is possession of stolen property," he says. "If a cop wants to be an asshole, they will charge you for it."

Cargo secured, Steve fires up the boom box affixed to his handlebars and mounts up. We roll off down the empty street, blasting the B-52s' "Dance This Mess Around."

To most people, Steve's work is invisible. He goes places most of us never see, trying to acquire objects others are trying to get rid of. One person's trash is another's treasure, as the adage goes. But when it comes to metal trash, that other person has to be motivated enough to put in some work and resourceful enough to find a market for that treasure.

There are scrappers like Steve in every city and town in North America. They like to talk to each other online, swap tips, show off big scores. They use smartphone apps like iScrap that show real-time metal prices and locations of nearby buyers. A surprisingly large number of them run YouTube channels, to which they post lovingly detailed videos of their excursions through back alleys and dumpsters. Some post tutorials on, say, how to extract the motor from a golf cart. Really popular channels, such as Mike The Scavenger's, rack up tens of millions of views and sell ads and T-shirts. The people who work in this shadow industry are often called scavengers, binners, bums, dumpster divers, or, to use Steve's preferred term, scrappers. A more precise label would be freelance metal recyclers.

Everyone loves the idea of recycling. It seems like such a self-evidently righteous thing to do: Use old newspapers to make new ones and save a tree! The version of recycling that most of us are familiar with makes it seem easy. You put your empty bottles, cans, and used paper in plastic bins, put them outside with your trash, and a truck comes along to haul it all away. All very tidy. Makes you feel like you're doing your part. So, when environmentalists say the way to avoid mining critical metals is by increasing recycling, it sounds intuitively right. Most people don't ask further questions.

That's certainly how I felt when I started researching this book. But I came to learn that the inescapable principle also applies to recycling:

Nothing comes without a cost. Recycling metal is, on balance, easier on the environment than mining. But it is also far more difficult, dirty, and dangerous than most people realize.

Though it's far off the radar of most media, metal recycling is a globe-spanning, multibillion-dollar industry. It involves heavy machinery, trucks, smoke-belching smelters, and cargo ships hauling millions of tons of cast-off products across the oceans. It involves thousands of scrapyard workers, as well as legions of independents like Steve, trolling through dumpsters and junk heaps all over the world. It involves people in developing countries with no safety equipment, or sometimes even shoes, smashing apart old motors and air conditioners with hand tools, exposed to the toxins they release.

In the United States, the scrap-metal trade dates back centuries. Alumni include (once again) Paul Revere, who melted down his neighbors' cast-off metal in his silversmith shop. The industry grew along with the country's economy. Today, scrap metal is a $40-billion-a-year industry employing tens of thousands of people. Three quarters of all the lead, half the iron and steel, and one third of all the copper America uses in a typical year come from recycled scrap.

Millions of tons of metal, however, are simply junked. Not because no one wants them, but because recycling them is difficult and expensive. It is often cheaper to dig up fresh, virgin metal than to reuse stuff that's already above ground.

How can that be? Think about the global supply chains that deliver a product like, say, an electric fan to your home. The chain begins with the extraction of raw materials in various parts of the world—aluminum for the blades, copper for the motor, nickel to keep the components from rusting. The ores that contain those elements are then transported to refineries to be processed into purified metals. Those metals are then sent to a manufacturing plant to be molded into parts. Those parts are then sent to a factory where they are combined with non-metallic parts and assembled into electric fans. Those fans are then shipped to stores and

Amazon warehouses all over the world. Finally, from these nodes where they are concentrated at the end of the supply chain, those fans are scattered far and wide to thousands of homes and workplaces.

To recycle the metals in those objects, you essentially need to run that process backwards. You need a *reverse* supply chain. Just as the regular supply chain winds through many countries and industrial facilities, so does the reverse supply chain.

The first link in the regular metal supply chain is the mine from which the metal originates. But as we've seen, that metal doesn't come out of the ground in ready-to-use form. It comes out as ore, a collection of rock and minerals that must be crushed, smelted, and run through a series of chemical and metallurgical treatments. At each step, extraneous material is stripped away, and valuable constituent materials are separated out.

The reverse supply chain is like a bizarro mirror version. The raw materials at the start of this backward chain are finished products: hot water heaters, car bodies, coffee makers, computers. That's why metal recycling is sometimes called "urban mining." Like ore, those products contain valuable metals that are mixed with, bonded to, or dissolved into unwanted junk material. Just as you need to separate out rare earths from bastnaesite ore, you need to separate out the copper embedded in a lawnmower from all the non-copper parts. To do that, you need to disassemble the lawnmower into its components, then break those components down to separate out discrete materials. Then, those materials need to be melted down, purified, and reconstituted as fresh metal, finally ready to be recycled into a whole new product. But first, you need to get your hands on those old, worn-out products, the raw materials of recycling.

In many ways, Steve is a unique character. But the key challenge he has to deal with is the same one faced by all actors at the beginning of the reverse supply chain: how to efficiently gather enough scrap to make the effort financially worthwhile.

That's difficult, because the final link in the regular supply chain isn't a link at all. It's a point of dispersal, like the end of a rope fraying into a thousand strands. Products arrive at stores or delivery warehouses in concentrated, orderly truckloads. But they leave as hundreds or thousands of individual products, each scattering out to a different office or home, spreading randomly across the land like dandelion fluff. To get them all back into the reverse supply chain, somehow all those atomized products have to be gathered back up again.

Big scrap dealers buy most of their feedstock in bulk—leftovers from construction sites and industrial facilities, for instance, or junk from demolitions. But that leaves out the countless tons of miscellaneous scrap scattered throughout millions of offices, small businesses, and homes. The first obstacle to collecting this metal is right outside your house—or rather, it isn't outside your house: There's no bin to put scrap metal in the way there is for paper and glass. Metal recyclers don't come to you. You have to go to them.

Where I live, in Vancouver, you can take most kinds of trash to a city-run depot optimistically called the Zero Waste Centre. I had never been there before I started working on this book. I decided to investigate (and to get rid of some junk).

Step one was gathering up household jetsam that couldn't go in the trash or regular recycling bins. There was plenty: a rusty folding chair, strings of old Christmas lights, dead cell phones, bits and pieces of plumbing left over from a bathroom remodel, worn-out pots and pans. I crammed it all into my car and drove out to the Centre, a large, fenced-in expanse of asphalt at the city's southern edge. The place is divided and subdivided into dozens of collection areas for a bewilderingly extensive list of reusable materials. The white packaging foam that coffee cups are made of is collected in one spot, the colored foam of pool noodles in another, and the spongy foam used to stuff cushions in yet another. Smoke and carbon-monoxide detectors each have their own collection bins, as do coin-size batteries, rechargeable batteries, and lead-acid batteries. So too with glass bottles, drinking glasses, and window glass. Each of these objects and materials has to be sorted and separated, because each requires a specific, unique recycling process.

It was up to me to sort all my various items and lug them to their respective receiving areas. (One of the few materials the facility does not accept is hard plastic, including a badly cracked blue recycling bin I wanted to get rid of. Yes: The recycling center would not recycle my recycling bin.) I did my best to be thorough, but I'm pretty sure that, at the bottom of the box of Christmas lights, there were also some nuts and bolts and maybe a few batteries that I just didn't bother to dig out. At some point, someone will have to sift through all of the junk I left (and everyone else's) more carefully. But that wasn't my problem. I'd done my part. To be honest, I was getting a bit fed up with the whole exercise, anyway. By the time I was done, it had taken a couple of hours, and all I had to show for my trouble was a marginally tidier house and a fleeting buzz of virtuousness.

People recycle metal this way every day—but, in the big picture, not many do. Why would they? It's a thankless chore and involves lifting and hauling and spending time in a junkyard, none of which are great incentives for most people. You know what's a better incentive? Money.

J. B. Straubel, a cofounder of Tesla who now heads Redwood Materials, a battery-recycling company, likes to say that "the largest lithium mine could be in the junk drawers of America." All those outdated phones, old charging cables, adapters, and other electronic bric-a-brac cluttering up your closets, to say nothing of the busted microwave in your garage or that rusty barbecue in the backyard, collectively contain enormous amounts of metals. In rich countries, these castoffs typically sit around unused or get thrown in the trash. In developing countries, however, where people are less inclined to just toss something that still has resale value, door-to-door scrap collecting is a significant industry.

Across Africa, Asia, and Latin America, millions of people are at work every day, shuffling through piles of what others have deemed worthless and pulling out the materials that in fact have value to all of us.

These waste pickers, as they're often called, gather discarded plastic, cardboard, glass, cloth, and metals, collecting volumes too small for big companies to bother with, and feeding them into that reverse supply chain. They provide a tremendous service by keeping junk out of landfills and reducing the need to extract virgin raw materials of all sorts. Though everyone loves the idea of recycling, most of the people actually doing the work are little noticed, badly paid, and wholly underappreciated. (Some of them are also children.)

In some developing countries, scrappers and waste pickers have organized into unions and won government support for their work. Essentially, they have been recognized as entrepreneurs performing a valuable service. In dozens of cities in Colombia, for instance, the government pays waste pickers a slice of the municipal trash fees to supplement the incomes they make selling recyclables. Buenos Aires, Argentina, has a similar scheme and provides space for pickers to sort their finds. In Pune, India, thousands of door-to-door waste pickers who belong to a worker-owned cooperative are paid small fees by the households they service—including in slums that regular city services ignore. They handle more than a thousand tons of waste every day. Pune has one of the highest recycling rates in the country.

Steve is the unorganized, non-government-supported Canadian counterpart to these workers. His father was a mechanic who was often called out to fix power lines connecting small towns in British Columbia's wild, mountainous interior. As a boy, Steve sometimes tagged along, flying out in helicopters and staying in remote work camps where the men drove ex-military vehicles and took potshots at bears scavenging in their trash dump. He went a little wild as a teenager. He was arrested a couple of times for underage drinking and was sent to an outdoor camp for juvenile offenders. "What kind of punishment is that? I loved it!" he says. At eighteen, he got his girlfriend pregnant, quit high school, and went to work in a chicken-rendering plant, cleaning gore off the cutters and grinders at day's end. Steve and his girlfriend split up when their son was about one, and Steve went to live in his parents' basement, but he and his dad didn't get along, so he moved on. He made good money as a firefighter in the interior for a couple of summers, and, in between, bounced around between family, friends, and a hostel. Eventually, he got a job with one of the local courier companies, delivering documents by bicycle. That suited him well. "I liked the biking," he says. "I liked the feverish pace." It was a good few years. He moved into steadier work as a dispatcher and spent his weekends mountain-bike racing with buddies.

But the document-delivery business was in terminal decline. First the fax machine cut into it, then the internet all but killed it off. "My job doesn't exist anymore," Steve says. "I went from three grand a month as a dispatcher to $250 on welfare." Steve found himself back in the hostel and lining up for day labor at a local employment center. He started picking up empty bottles and cans and returning them for the deposit to make a few extra bucks. It turned out to be more lucrative than he expected. "Before I know it, I've got ten dollars in empties, and me and my cat can eat for two days," he recalls. As he worked the streets, Steve learned more about what types of trash could be sold and where to find them and soon realized scrap metal was the niche for him. He scouted out his trapline and has been working it, more or less, ever since.

Somewhere around 2008, Steve was robbed in a parking lot by a guy who hit him in the head, hard. He was hospitalized with a fractured skull and a detached retina. "There's a few years that are a bit cloudy because of the head injury," he says, "and some drug use." He was arrested and jailed a few times. He wound up living under an overpass for several years. He preferred it to a welfare hotel. "I tried one of those places, and I couldn't get any sleep. I could feel mice running over my body. My clothes were infested with bugs. I got a better night's sleep under the bridge."

When I meet him in 2022, though, Steve has a small apartment in Vancouver's Chinatown and is feeling upbeat. "I'm gonna make up for lost time. I've been given second chances and third chances and opportunities that were blown. So, I want to make the most of it," he says. He likes scrapping—the variety, the people he's gotten to know, the freedom. "I've been doing this for twenty years. I'm my own boss. If I have a shitty day, it's nobody's fault but mine."

He's a professional, Steve is, and he takes his job seriously. He feels that too many scrappers don't bother with the lower-value metals like aluminum. "I find most of them now are not willing to do the work. They just want the easy copper, the easy brass, the stuff that's right on top of the bin. They're just looking for something they can sell quickly."

Like any good entrepreneur, Steve is always looking for new markets. He's been collecting obsolete circuit boards tossed out by an elevator-repair outfit for years. No one in town buys scrap circuit boards, but Steve knows they contain small amounts of gold. One of these days, he'll get around to watching the many YouTube videos that demonstrate how to extract gold from circuit boards and figure out how to do it himself.

After spending several days running the trapline with Steve and hanging around scrapyards, I find myself looking at everything around me in a new way. I'm suddenly aware of all the metal everywhere: in rain gutters, bicycle chains, water pipes, trash cans, streetlight poles, and the cables inside those poles, fat with precious copper. Metal is like concrete, a material that is everywhere but that we don't see, either literally because it's concealed inside something else, or figuratively because we just take its presence for granted. But it nonetheless gives shape to our surroundings, underlies the physical structure of our urban environments. Metal permeates our personal lives, too, in the tip of a ballpoint pen, in the button and rivets of a pair of jeans, in the handle of a dresser drawer. And every bit of it is worth some tiny amount of money.

I find myself wondering how much I could get for my old ski poles or tattered folding chairs. It's like having a sort of superpower: I can see the value in things that other people can't. What to most people is just unsightly, broken, useless debris, I now see as money.

But thinking like this also makes me aware of how difficult it is to actually turn junk into money. That old showerhead in the garage has some sellable zinc and aluminum in it, but who is going to break it apart and separate the metals from the plastics they're attached to? Not me. It doesn't seem worth the trouble. As I learned from my trip to the Zero Waste Centre, anyone can recycle metal. But in rich countries, only a few of us do.

Further on along the trapline, Steve and I cruise into a cul-de-sac sandwiched between two abandoned, boarded-up buildings. Steve gives me a quick rundown of the small manufacturing businesses that have come and gone through them. He is probably the only person alive with such a granular historical knowledge of the area. These days, the cul-de-sac is a popular spot to dump junk illegally. It is strewn with trash of every description—paper, cardboard, bits and pieces of plastic and metal, and an abandoned stove and a fridge sitting enticingly on the sidewalk. If the stove has an induction cooktop, Steve explains, the burners will include a copper coil. "A nice, thick, braided wire. I'd rip that out in a second!" he enthuses. But someone else beat him to it. The appliances have already been stripped, their power cords snipped off and innards extracted. The stove burners turn out to be zinc.

Someone will come by during the night with a truck for the steel and aluminum husks, Steve assures me, even though they're worth only a few bucks each. Steve has semi-exclusive rights to some of the dumpsters on his trapline thanks to his long-standing relationships with the buildings' tenants, but this spot is no-man's-land. First come, first served is generally the etiquette, he explains. "Though I did have a guy come at me with a steel bar three weeks ago," he adds, as an afterthought.

Scrapping can be shockingly dangerous. One minute you're rooting through a heap of precariously piled metal junk, the next your foot gets crushed by a dislodged air conditioner, or your finger is sliced off on a jagged piece of steel. Hundreds of American scrapyard workers have died on the job since 2003. The main killers are accidents involving heavy machinery, but workers sometimes run into other hazards. Two people died in 2014 when an old mortar round blew up in an Illinois metal-recycling yard. The following year, a scrapyard worker in Arizona was using a torch to cut through a piece of metal that turned out to be a misplaced Air Force bomb; it exploded, killing him.

In places where poverty and warfare are prominent features of daily life, scrappers go after military debris deliberately. Eight children were blown to pieces in Sudan in 2019 while scouring for scrap near a military base. In Nigeria, Boko Haram insurgents have killed dozens of scrap-metal collectors. In Afghanistan, decades of war have strewn countless tons of metal across the landscape. *The New York Times* reported on one area where "the rolling hills, between jagged mountains, have turned into a congealed mass of discarded steel and hidden explosives," making the place "a scrapper's fever dream, a place where 15 pounds of discarded metal can be quickly harvested and sold for around a dollar." Dozens of those scrappers have been killed by unexploded mines and bombs.

MOST BUSINESSES THAT CALL THEMSELVES METAL RECYCLERS DON'T ACTUALLY turn old junk into new metal. They are primarily collectors, aggregators. They process the junk, to varying degrees, and sort it into piles big enough to sell on down the chain to someone who can take it to the next stage. This is exactly what Steve does, on the scale of a single person. He gets some of his material for free from dumpsters but also buys a bit here and there. One day, he showed me a bale of wire that he'd bought from a movie-set decorator who was retiring and needed to get rid of all the gear he'd accumulated. He offered Steve the wire at a cut rate, which Steve accepted once the decorator cleaned off the colored tape stuck all over it. That left Steve to strip the rubber coating off with a box cutter. "So, he's doing some of the work, I'm doing most of the work, but we're both gonna come away with something," Steve says without rancor.

It's rare that Steve finds enough pure metal to make it worthwhile to take it straight to the scrapyard. Most of what he finds can be made more valuable if he processes it—putting in the effort to strip the wires, break off the brass valves, and otherwise sort and separate out the metals as much as he can by himself—rather than selling the scrap as he found it. Sometimes he breaks down scrap in his tiny apartment, where he keeps a set of tools. He also rents a small storage locker, where he stashes overflow scrap. He uses it as a disassembly room, as well. On rainy days, he'll hole up there, crank some heavy metal, and strip wire all day long.

A few days after we run his trapline, Steve bikes up to Capital Salvage, a tiny scrapyard tucked in the middle of a nondescript light-industrial block on Vancouver's east side. The place is hidden from the street by a concrete-block wall. You have to go down an alley to find the entrance, a barbed-wire-topped steel gate. Inside is a maelstrom of cast-off and worn-out manufactured goods. It's as though someone had dragged a Wile E. Coyote magnet through a random city block to pull out every metal-bearing object, then scattered them around this asphalt courtyard. There are desk chairs, vacuum cleaners, a kid's scooter, stacks of aluminum siding, a box of rusty saw blades, ice skates, window frames, a spatula. There are radiators, microwaves, floor fans, ceiling fans, lawn chairs, car-wheel rims, beer kegs, a bike rack, and electric wire in every size and color. The angry buzz of power saws, the grumbling of a forklift motor, and the clatter of bits of metal getting tossed from one pile to another fills the air. From the roof of the cutting shed, an astronaut someone made out of old aluminum ducts, a fishbowl, and cables waves a gloved hand.

Steve parks his bike by the gate and gets to work. His trailer is decorated today with a little Ukrainian flag he picked up somewhere and is loaded with about one hundred pounds of neatly folded brass sheets, some stainless steel pipes, and a coil of copper wire, all tied down with twine and old inner tubes. He says hi to Jen Dimant, the yard co-owner, who is staffing the gate, and gets busy separating out his haul into the big plastic bins set out for this purpose.

When everything is sorted, he takes the bins over to a set of outdoor scales. Two Capital Salvage workers note the weights of each different type of metal and print out a paper slip for him. Steve takes the paper to a cash dispenser, which is tucked in between two old vending machines and decorated with a sticker bearing the number of the city's Overdose Outreach Team. Steve feeds his slip into the machine and it spits out $370 in cash. Only then does he take a close look at the receipt. Some of his copper was graded as #1, instead of the higher-purity #2, which would have fetched a few more dollars. "Only $3.50 a pound? Why is that copper number one?" he wonders aloud. "I'm gonna call him on that." He ambles off to argue with the scale guys. "Why get rich off the little guy?" he mutters.

Steve doesn't have a lot of bargaining power, though. If you have a car or truck, you can sell your junk to any scrapyard in town, but Capital Salvage is the only one in Vancouver that accepts walk-ins like Steve. Scrapyards have a shady reputation. They are often suspected of buying stolen metal—and some do, knowingly or otherwise. In Vancouver, as in many other places, waves of copper and other metal theft have put pressure on the yards to clean up their act. (At one point, the nation of Kenya shut down its entire scrap industry in a bid to curb an epidemic of metal thievery that was crippling the country's power grid.)

"A lot of the scrapyards, their solution to clean up their image was, 'no more walk-ins, cause they're homeless and drug addicts,'" explains Dov Dimant, Jen's husband, a burly guy with an unruly black beard wearing well-worn overalls. "It's easy to point the finger and blame homelessness and drug addiction and stuff like that. Where, in reality, people committing these crimes aren't pushing shopping carts. They have pickup trucks. They're normal-looking people. The people pushing shopping carts, they're generally just digging through dumpsters, walking through alleys, stuff like that."

"I didn't have the luxury of turning away all that business, nor did I think it was right. It was very discriminatory," he continues. "So, I don't do that. As long as you have proper ID, and you're not causing any trouble, come on in. If you have one pound, if you have ten thousand pounds, doesn't matter."

Capital Salvage has even partnered with a local Vancouver social-service organization to print identity cards for homeless scrappers who lack the ID required by law to sell scrap. "If we can make recycling metal more accessible and get more income for the people that are doing the job, who really get paid the least, that's good for everybody," says Amber Morgan, who created the program.

Dov inherited Capital Salvage from his parents and now runs it with Jen and about twenty employees. When he was a kid, there were half a dozen scrapyards within a few blocks of their operation. Now, most have closed or moved to the city's fringes, pushed out by soaring property prices and concomitant gentrification. The block Capital Salvage sits on used to be all light industry and warehouses. Now, there's a chic microbrewery right across the street. That type of neighbor tends to frown on having a noisy, dirty business in its midst. It's an echo of what happened to the mining and manufacturing industries, so much of which were pushed overseas because Americans didn't want pollution and unsightly facilities in their own backyards.

Dov is Jewish, which is surprisingly common for scrapyard owners. It's a demographic quirk that dates back to the wave of Eastern European Jewish immigrants that poured into the United States and Canada around the beginning of the twentieth century. Most of them were broke, didn't speak English, and were barred from all kinds of jobs. Of necessity, many of them turned to niche trades, especially ones that didn't require much startup capital, like rag-picking.

"Since the work was dirty, dangerous, and low status, few natives with

other prospects chose to perform it for any length of time. The low starting costs, combined with a lack of competition from established natives,
made it possible for immigrants to gain footholds in the trade," writes
the historian Carl Zimring, as quoted by Adam Minter in *Junkyard Planet*,
an examination of the global scrap trade. Jewish peddlers wandered the
streets with pushcarts, collecting old clothes and other castoffs that they
could resell. Over time, some of them branched out into glass, paper, and
scrap metal. They built thriving businesses out of trash.

At Capital Salvage, almost all the metal comes to them, brought to
the alley gate by plumbers and electricians with leftover bits of pipe or
wire, and by scrappers like Steve. Some comes from random people who
might be cleaning out their basement, but not a lot. Jen says people call
them all the time asking how much they can get for their household junk.
"When we tell them the prices, they say, 'Ah, I'll just throw it out. It's not
worth it.'" Capital Salvage handles the first few steps of processing the
product that comes in, using machines to strip wire and gas-powered
saws to cut apart appliances and furniture. But they can't process really
big items, and they don't have the space to store large volumes of anything. So they sell their semiprocessed, semisorted scrap on to the next
link in the chain—a bigger scrapyard, like ABC Recycling in the Vancouver suburb of Burnaby.

At this point, a river might be a better metaphor than a chain. Steve's
scavenged junk is like a tiny trickle that feeds into the larger stream of
Capital Salvage, which, in turn, feeds into the tributary of ABC, which
will feed it into a global river of reclaimed metal, rushing not toward the
ocean but a blast furnace.

ABC Recycling grew from the efforts of Joseph Yochlowitz, a Jewish
immigrant who came to Vancouver from Poland in 1912 and started
peddling scrap with a horse and wagon. Today, it's run by the fourth generation of his descendants, a $170 million business with more than two
hundred employees working in nine locations. They are fed by many
streams. The company's ten-acre operation in Burnaby takes in scrap from

demolition companies, manufacturing plants, and smaller brokers like Capital Salvage. "You name it. We deal with steel distributors, mines, the movie industry, equipment suppliers, waste companies, demolition companies, plumbers, electricians, HVAC centers," explains Randy Kahlon, vice president for sales. Some scrap gets brought to them, some they go out and find. Randy has a team of people always on the hunt for new sources, bidding against other yards for big scores. Metal prices are notoriously volatile, so buyers have to be quick on their feet. When I visited the ABC yard one gloomy spring day, it was only a few weeks after Russia had invaded Ukraine, and prices were shooting up.

The yard itself could hardly look more apocalyptic. It's like a vast charnel house after the robot Armageddon, strewn with towering piles of twisted, blackened, rusting metal. Humans can be glimpsed here and there atop scurrying forklifts, but most of what's moving around are enormous machines built to destroy. A machine that looks like a titanic, angry turtle methodically chomps the screeching, groaning steel and aluminum skeleton of a railroad car into pieces. Compressors squash old cars down into lumpy metal pancakes. Cranes wielding giant magnets and four-clawed grapples move snarls of chopped-up rebar from here to there, dropping them with a raucous clamor. There are heaps of gleaming copper and aluminum wire, hillocks of steel rebar taller than a house. There are crates filled with brass knobs or zinc plumbing fixtures and stacks of effulgent, hay-bale-like cubes of shredded aluminum.

Kahlon, like just about everyone else in the scrap business, is well aware of the critical-metals boom and is positioning the company to take advantage of it. "The world is going electric, so the world needs more of this stuff, right?" he says. "All the prices have been skyrocketing for the last couple of years, and it's purely because there's not enough supply. That is going to be the major problem. There's only two ways to get it. You either mine it, or urban mine it," he continues. "The demand for recyclers and what we do is only going to intensify."

As Kahlon will acknowledge, however, his shop does not actually recycle. They're another link in the chain, another stretch of river. They take in junk, shear or slice or smash it down to manageable sizes, sort it by type, bale it up, and move it along. Only much further downstream will it be melted down, recast, and U-turned back into the supply chain. Some of ABC's product gets taken by rail or truck to steel mills in Washington State, but most of it gets loaded into containers and put on ships. I bet you can guess where most of those ships go.

Since the 1990s, China has been gorging on metals to feed its explosive industrial expansion. Today, it is the world's leading consumer of steel, copper, aluminum, and many other metals. But China doesn't have a lot of its own metal resources. So, to help fill the gap, it turned, decades ago, to imported scrap. Its timing was fortuitous, as Minter explains in *Junkyard Planet*. "The last of the American factories devoted to refining copper from scrap metals shut down in 2000 due to the high cost of complying with environmental regulations," Minter writes. "Partly as a result, China, which barely had a copper refining business in 1980, now has the world's biggest. Not only that, it has some of the best, most technologically advanced (and environmentally secure) copper refineries in the world." Minter visited one town that salvaged millions of pounds of copper each year from a single source: imported Christmas lights. Competition for feedstock is fierce. "In China, lavish dinners, often concluded with prostitutes, are often just the base entry requirement if you want to even talk with certain factory bosses about their scrap," writes Minter.

Chinese entrepreneurs quickly realized there was a lot of money to be made wrangling not just scrap but all of the West's garbage. By the early 2000s, it was the world's leading recycler of everything—cardboard, plastic, metal, anything that industrious hustlers could squeeze a little extra value out of. One of the world's richest women, a billionaire Chinese entrepreneur named Zhang Yin, made her fortune importing wastepaper from the United States and recycling it into cardboard boxes,

boxes in which Chinese goods are packaged for export back to North America.

Today, China's metal-recycling industry pulls in around $60 billion per year and employs a quarter of a million people. "They [have] become, quietly, the Scrapyard to the World, a place where wealthy countries sent the stuff that they couldn't or wouldn't recycle themselves; a place where former farmers took that stuff, made it into new stuff, and resold it to the same countries that had exported it in the first place," writes Minter.

Wherever it's done, recycling metal has environmental benefits that help us all. Every ton of metal that gets recycled is a ton that doesn't have to be dug out of the ground, with all the destruction that entails. Recycling doesn't require clear-cutting forests or tapping desert aquifers, let alone digging up the seafloor, and it helps keep junk out of landfills.

But, as always, there are costs.

Grinding and shredding up all that scrap throws off particulate matter—metal dust, essentially—that can travel away on a breeze and end up in the lungs of people living nearby. The intense heat involved can also vaporize whatever plastics, paints, sealants, and other cruft is in the scrap, creating more airborne toxins that can infect the water and air of surrounding communities. Scrapyards across America are often found to be emitting toxic materials.

Melting down scrap also requires huge, powerful, extremely hot furnaces. In China and elsewhere, much of the energy to create that heat comes from carbon-spewing coal and natural-gas plants. All told, the carbon emitted per ton of recycled metal is generally less than that of a ton mined from the ground. But it's far from zero.

As a result of all these collateral effects, China got fed up with being the world's recycling bin. Beginning around 2017, Beijing blocked the import of most solid waste, including scrap. But with the demand for critical metals sharpening, China has reopened its doors to some forms of scrap copper and aluminum, as well as iron and steel. Metal trash is just too important, China decided. The scrap industries in India and other countries are also growing and vying for the rich world's castoffs. Several countries are making efforts to keep their scrap at home, taxing or banning the export of used metal. In a world hungry for critical metals, scrap metal is increasingly recognized as treasure. The industry keeps growing, all around the world, creating opportunities in the least likely places.

CHAPTER 9

High-Tech Trash

meet Baba Anwar in a crowded, chaotic market in the city of Lagos, Nigeria. Maybe all of five feet tall, he is wearing plastic flip-flops, shorts, and a filthy T-shirt that says "Surf Los Angeles" and clutching a printed circuit board that had once been inside a laptop computer. Anwar, who looks fifteen or sixteen but claims to be in his early twenties, tells me he found it in a trash bin. That's his job, scrounging for discarded electronics in Ikeja Computer Village, one of the world's biggest and most hectic markets for used, repaired, refurbished, and occasionally counterfeited electronic products.

The market fills blocks and blocks of narrow streets, all swarming with people jostling for access to hundreds of tiny stalls and storefronts offering to sell, repair, or accessorize digital machinery—laptops, printers, cell phones, hard drives, wireless routers, and every variety of adapter and cable needed to run them. The cacophony of a thousand open-air negotiations is underlaid with the rumbling of diesel generators, the smell of their exhaust mixing with the aroma of fried foods hawked by sidewalk vendors. Determined motorcyclists and women in brightly colored dresses carrying trays of little buns on their heads thread their way through the crowds.

It's not the easiest place for an in-depth conversation. But, with the help of Bukola Adebayo, a local journalist, I gather that Anwar came to Lagos about a year earlier looking for a better future than his deeply impoverished home state of Kano had to offer. "No money at home," he explains. In Lagos, a pandemoniac megalopolis of more than fifteen million, he shares a room with a couple of friends from back home, all of them e-waste scrappers. On a good day of scavenging electronics at Ikeja, he can make as much as $22. And on a bad day? "Nothing."

There are thousands of Nigerians making their living the same way Anwar does—by recycling electronic waste. E-waste, as it's commonly known, is a broad category of detritus that includes just about anything with a plug or a battery, and the components thereof, that has been thrown away. That includes the stream of digital devices—computers, phones, game controllers—that all of us are using and tossing out in ever-growing volumes. The world already generates more than fifty-three million tons of e-waste every year. That's enough to fill "a million 18-wheel trucks stretching from New York to Bangkok and back," according to a UN report. The total is projected to rise to seventy-five million tons by 2030.

Digital e-waste is a uniquely troublesome form of trash, because if you just dump those gizmos in a landfill, toxic chemicals can leach out of them into the soil and water. In rich countries, most people have no easy way to recycle their old Samsung Galaxy phones, Xbox controllers, and other devices, so tons of them end up gathering dust in junk drawers and garages. The United Nations estimates that, worldwide, only 17 percent of all e-waste is collected and recycled. The rest gets dumped, burned, or just forgotten about. The question of how to deal with it all, and who will pay what price, is growing more urgent—particularly since e-waste is also an important potential source of critical metals. Digital gadgets contain a whole gamut, from the copper in their cables to the lithium, cobalt, and nickel in their batteries. All told, that means the world is wasting close to $60 billion worth of metals in e-waste every year.

The problem is that digital devices contain only a small amount—sometimes a *really* small amount—of each metal. The scrap value of a single phone, earbud, or Fitbit is pretty low. But in poor countries like Nigeria, there are lots of people willing to put in the time and effort required to recover some of that value. Their role is much like Steve Nelson's in Vancouver—individual little streams disemboguing scrap into the global river. The difference is that there are many, many more of them, working for even less money. There are no official statistics, but research indicates that there are tens of thousands of e-waste scavengers in Nigeria alone. Some go door-to-door in residential neighborhoods with pushcarts, offering to take away or even buy people's unwanted electronic goodies. Some, like Anwar, work the secondhand electronics markets, buying bits and pieces of broken gear from small businesses or rescuing them from the trash. Many of them earn less than the international poverty line, currently about US $2.15 per day.

I ask Anwar where he is planning to take that circuit board. "To TJ," he answers, as if I had asked him what color the sky is.

From a dingy concrete building overlooking the Ikeja market from across a roaring four-lane road, Tijjani Abubakar—known across Lagos as TJ—runs a thriving business turning unwanted electronics into cash. His third-floor office is a charnel house of dead cell phones. At one end of the long, crowded room, two skinny young men armed with screwdrivers pull mobile after mobile out of a sack and crack them open like walnuts. Their practiced fingers pull out the green, printed circuit boards inside and toss them with a clatter on a heap at their feet. There are thousands of the little boards glittering flatly under the overbright LED ceiling lights. More young men scattered around the floor on plastic stools sort the boards into smaller piles, sifting out those with the most valuable chips. The air is thick with sweat despite the open windows. The room is full of low-voiced chatter, the clack-clink of tumbling circuit boards, and the relentless honking and engine noise of the traffic outside.

At a scuffed wooden desk at the far end of the room sits Abubakar himself. He is a big man with a steady demeanor, lordly in an embroidered brown caftan, red cap, and crisp beard. I wait my turn to speak with him as he fields calls and WhatsApp messages on three different phones and a laptop while negotiating a deal with a couple of visiting traders over an unlabeled bottle of something.

Abubakar has been in the e-waste business for nearly twenty years. He's originally from Kano, like Anwar and many others involved in the Lagos scrap trade—migrants who've found the same economic niche that Jews did in North America a century ago. His father sold clothes—"not a rich man," says Abubakar in his even baritone. Abubakar earned a degree in business from a local university and made his way to Lagos in search of opportunity. A friend introduced him into the e-waste trade.

"We started small, small, small, small," he says. Back then, it was easy to get a foothold in the business. Scrap was cheap or even free, because so few people were interested in buying it. At first, he dealt only with other Nigerians. But as the e-scrap supply mushroomed, so did the competition. All kinds of foreign e-waste buyers—from India, Lebanon, and, above all, China—have flocked to Nigeria, many of them with deeper pockets than Abubakar.

No one gives away scrap anymore. "Now everybody knows the prices," says Abubakar. Electronics dealers in the Ikeja market want to be paid for their castoffs. Still, his business has flourished. By now, he exports several shipping containers of e-waste every month to buyers in China and Europe. He has grown wealthy enough to donate textbooks, meals, and even live cows to families back in Kano. Dead cell phones converted into education and food. Trash into possibilities.

Abubakar buys and sells all manner of e-waste, but he specializes in mobile phones. It's a solid twenty-first-century business. All over the developing world, mobile phones have become as common as T-shirts. There is just shy of one registered mobile account for every single one of Nigeria's 220 million people. "Everybody has a phone. Not everybody has a computer," he says. "What do I see here?" he asks, waving a hand at the room full of workers and gutted mobiles. "I'm the only one with a computer. I don't know whether any of these people have a computer. But I know all of them have a phone."

Those phones, like all phones, eventually wear out, break down, or get tossed by owners eager for a newer model. Worldwide, in 2022, an estimated 5.3 *billion* mobile phones were thrown out. If you stacked them all up, the pile would reach one eighth of the way to the Moon.

Abubakar deploys a vast network of buyers and pickers to seek out discarded phones all over Nigeria, as well as in neighboring countries and, occasionally, as far away as France. The phones and other e-waste arrive by truck, by train, and in sacks carried by workers like Anwar. All day long, they tramp into the muddy courtyard next to Abubakar's building, where more workers weigh their sacks of electronic scrap on a digital scale. The bigger items, like computer towers and wireless routers, are stashed in a tin-roofed shed. On the day I visited, three men were hunched on a concrete pad by the shed, smashing phones apart with hammers and prying out their circuit boards like meat from a crab shell.

At the beginning of the phones' lives, these precisely engineered products were manufactured under ultra-clean conditions in sophisticated, high-tech factories; now, at their end, they are being torn apart by hand on a grimy concrete pad. It's an oddly disturbing sight, perhaps because phones are more intimately connected to us than, say, toasters. Every one of these phones must have been held for hours in someone's hands, sat for days in someone's pocket or purse, a constant digital companion. There's something jarring about seeing their viscera ripped out like this.

All told, Abubakar says, he has about five thousand people working for him, bringing in millions of phones each year. I express polite skepticism at this figure. Without changing his expression, Abubakar rises from his desk and gestures for me to follow him through a door in the back of the office. He leads me into a warren of rooms, all of them filled either with enormous sacks stuffed with phones, people cracking and sorting phones, or bales of circuit boards ready for shipping. The urban-mining label fits. It's an inversion of the industrial order: Workers digging through piles of manufactured products to produce raw materials, rather than the other way around.

There are thousands and thousands of Sonys, Motorolas, Samsungs, HTCs, iPhones, and a hodgepodge of Chinese brands, most of them low-end smartphones. Abubakar's workers extract every penny of value from each. Just like scrappers in North America, Abubaker makes more money if he sorts and processes his wares before selling them. Different brands and models are sought by different buyers and command different prices, he explains. "See these, these are only Android phones," Abubakar says, gesturing at an open sack full of handsets, "eight gigs, sixteen gigs, all together. We need to sort them out."

The main components buyers want are printed circuit boards, the thin panels of (usually) green plastic or fiberglass found in everything from kids' toys to medical devices. The boards are etched with copper pathways that carry signals between the soldered-on chips, capacitors, and other parts.

Abubakar's workers break the chips off the boards for further assessing; if they still work, they can be sold separately for use in refurbished phones. On his own phone, Abubakar pulls up an eye-crossing list of hundreds of chip serial numbers and their corresponding prices. He shows me a lunch-bag-sized sack of Android chips. The serial numbers printed on them are so tiny I can barely make them out. "In dollars, this bag is worth around $35,000," he says. The cameras, which are just the little lens you see on the phone's back attached to a strip of metal foil inside it, can also be extracted and sold separately. A sack of those is also worth quite a bit. Abubakar keeps security cameras trained on his workers to make sure nobody pockets anything—cameras watching people extracting cameras. He fired someone the week before for stealing some chips.

It takes a sharp and experienced eye to properly separate out the boards and chips. "It's not a today-tomorrow job. It's a lot-of-time job," Abubakar says. In one of the rooms, a young Chinese woman is squatting on the floor amid the other sorters, the only non-African person I've seen amid the dozens in Abubakar's compound. He explains that he hired her away from a Chinese competitor for her special expertise about the value of certain phones, which he declines to explain. "There was something I didn't know very well, that she knows all about," he says. "It's something like a secret that the Chinese are hiding." Whatever her classified intelligence is, it's valuable. To keep her on board, Abubakar says he bought her a car.

None of these thousands—millions?—of phones were made in Nigeria. None will stay there, either. It's easy to pull a circuit board out of a phone or a laptop. Getting the metals out of the circuit board, however, is another story. Shredding or melting down a circuit board and separating out those tiny amounts of gold, copper, and everything else requires sophisticated and expensive equipment. There is not a single facility anywhere in Africa capable of performing this feat. So, Abubakar sells his wares to the small number of recyclers in China and Western Europe who have the proper equipment. In other words, he is getting rich selling trash from one of the world's poorest countries to some of the world's wealthiest. In the case of China, that also means he's selling the dead phone parts back to the country that probably manufactured them in the first place.

The problem of rich countries illegally "dumping" e-waste in poor ones, where environmental and labor standards are more lax, has received a lot of attention since the early 2000s, kicked off by activist groups like the Basel Action Network and news outlets such as *60 Minutes*. Dumping is still a problem, but, these days, there is probably at least as much e-waste moving in the opposite direction. From Bangalore to Jakarta, there is a growing tide of e-waste flowing from the developing world to the rich.

No one knows exactly how much e-waste travels between countries.

There's no standard definition of what the term even means, much less a global system for tracking it. But the United Nations estimates that about five million tons of e-waste, less than 10 percent of the global total, is moved from one country to another each year. Of that five million tons, almost two million are transported aboveboard, with proper permits and licenses. That leaves about three million tons that are shipped in an "uncontrolled" manner, some of which may get illegally dumped. That's a significant problem.

But the world has changed a lot in the last twenty years. Today, most of the e-waste in West Africa and other parts of the developing world is generated domestically. As Abubakar says, everyone has a cell phone. In developing countries, people buy most of those phones and other devices secondhand, from businesses that import used electronics from overseas. Most of the discarded computers and cell phones in Nigeria weren't smuggled in from some Western country that didn't want the hassle of disposing of them; they were imported as new or secondhand products, sold to local consumers, used, and *then* discarded.

Some of the five million tons of e-waste that goes from one country to another is not actually trash but used electronics that will be reused by less choosy consumers. Some of it, of course, *is* trash—broken electronic devices and fragmented parts. But even those castoffs are moving not only because someone in a rich country wants to get rid of them, but because someone in another country wants to buy them. Abubakar buys broken phones from France on the cheap, imports them to Nigeria, pays his workers to break out their circuit boards, chips, and cameras, and still makes a profit selling those segments to recyclers in China. It's a growing trade. Buyers from Chinese companies, as well as European outfits like Belgian recycling giant Umicore, are actively scouring the scrap markets of Africa and Asia in search of recyclable materials to ship to their home countries.

Meanwhile, most of the rich world's e-waste isn't getting recycled at all. In the United States and Europe, fewer than one in six dead mobile phones is recycled. Of all the electronic gear junked in Europe each year, only about one third winds up in an official recycling system. "It's surprising, especially in Europe, because we have fully functioning recycling systems," says Alexander Batteiger, an e-waste expert with the German development organization GIZ. Nobody in the rich world is going house-to-house, knocking on people's doors to offer them a couple of bucks for their old iPhone 6 or Bluetooth speaker. Sure, there are e-waste collection drives at schools and churches, and you can take old electronics to stores like Best Buy or public facilities like the Zero Waste Centre—but how many of us bother? Not many. Instead, countless millions of phones and tablets and blenders and microwaves are abandoned in American junk drawers, closets, and landfills.

Things are different in the developing world. In Nigeria, as in many poor countries, the recycling trade is overwhelmingly dominated by what is politely called the informal sector—people working without permits or licenses in markets of various shades of gray. Informal scavengers in cities like Lagos *do* go door-to-door buying household e-waste for pennies or accepting it for free—links in the reverse supply chain. If you are living on $2 a day, making ten cents from a discarded electric toothbrush is well worth your time. The result is that about 75 percent of Nigeria's e-waste gets collected for some kind of recycling. In nearby Ghana, the number is estimated to be as high as 95 percent.

There's plenty to celebrate in this cycle of trade. Businesses like Abubakar's are keeping potentially toxic trash from ending up in landfills, reducing the need for mining fresh metals, and creating thousands of jobs. That is not a trivial consideration in Nigeria, where nearly two thirds of the country's 220 million people live in poverty.

But neither is it the whole story.

An hour's drive from TJ's office, through the maelstrom of Lagos's notorious traffic, lies the Katangua dumpsite. It is a sprawling, teeming maze of tiny workshops, scrapyards, wrecking zones, and slums, loosely centered around a mountain of trash heaped at least twenty feet high.

The garbage colossus is surrounded by a fence of made of rusty fragments of corrugated tin held in place with bits of scrap wood. Plumes of thick black smoke wend upward from somewhere inside the fence. Atop the trash sit several shanties, where people presumably live.

The sheer squalor of Katangua is unfathomable. The ground underfoot is churned up mud inlaid with trampled-in plastic trash. Barefoot children wander between shacks made of cardboard, plywood, and plastic sheeting. Bukola and I pick our way around the bathtub-sized puddles, following men and women carrying sacks full of discarded metals, retreating to the roadside whenever trucks piled high with aluminum cans and heaps of scrap wallow past.

Practically every type of scrap metal and e-waste is recycled somewhere in Katangua, one way or another. The resourcefulness of the people who work there is as astonishing as their working conditions are appalling. We stop in at one yard where the owner, Mohammed Yusuf, is proud to show me his aluminum-recycling operation. Pickers bring him aluminum cans scavenged from all over Lagos, two or three tons every day. The front part of his yard is filled with heaps of sacks bulging with cans. At the back, there is a covered area with a rectangular hole about the size of a bathtub dug into the ground and lined with bricks. There's a smell around it reminiscent of rotting chicken. At night, Yusuf explains, his workers fill the hole with cans, melt them down with a gas-powered torch, then scoop up the molten metal with a long-handled ladle and pour it into molds. When it cools, the result is silvery, rectangular, two-kilogram ingots of more or less pure aluminum ready to be sold to a manufacturer who will turn that metal into new cans. The process generates intensely toxic fumes and dust, so his workers wear protective masks. What about the other people nearby, I ask? Yusuf nods. That's why they only melt the cans at night, he explains, when the people who live in the shacks around the yard are in bed.

In another area, a swarm of men caked in filth are breaking down the carcasses of old trucks by hand, whacking apart engine blocks, prying off bumpers, and pulling out shock absorbers with hammers, crowbars, and lengths of rebar. It's straight muscle versus metal. God only knows how much oil and engine fluid drains out into the mud in the process.

Nearby, a couple of young men in surf shorts sit in a little alcove in front of a shuttered stall, dismantling laptop computers with screwdrivers. The circuit boards go onto a good-sized heap next to them, the tiny cameras into a plastic bag. Those items will be sold into the reverse supply chain. The plastic components, however, will end up in the dump. A worse fate is in store for the copper-bearing cables.

Squeezing through a gap in the ragged fence, Bukola and I find ourselves in an open area at the base of the towering garbage pile. There, four lean young men are tending several small fires. They're burning the coatings off piles of wire to get to the copper inside them. The flames are quite beautiful, deep blues and greens licking up amid the orange. The smoke is thick and oily and reeks of incinerated plastic and rubber; it almost certainly carries highly toxic dioxins. The men are wearing shorts, T-shirts, and flip-flops—no respirators, no protective glasses, nothing remotely resembling safety gear.

Between the wire burning, the open-air aluminum smelting, and the car wrecking, I'm horrified by the thought of how thoroughly poisoned the whole area must be. A 2019 study by Nigerian researchers found that informal e-waste scrapping in Lagos released heavy metals and other contaminants into the soil, water, and air, with possible harmful effects "on the entire environment and in particular human health."

"Do you worry about breathing the smoke?" I ask the friendliest burner, a muscular, lightly bearded young man named Alabi Mohammed. He's wearing a tank top, his muddy feet in purple flip-flops. "This is what we are used to," he shrugs. "We don't know any other job. We don't have any other option." He's been living in the market since he was eight years old, he says. At the time we met, he was thirty-six.

There are other destructive e-waste recycling practices that I didn't see in Katangua. Printed circuit boards often contain precious metals like palladium, silver, and gold. That makes them a potentially very efficient source of metals. One ton of circuit boards can contain forty to eight hundred times the amount of gold found in a ton of ore, according to the US Environmental Protection Agency.

One way to extract that gold is to run the boards through a shredder and then ship the fragments to refineries, typically in Europe or Japan, where the gold is extracted using chemicals. "It's a precise, mostly clean method of recycling, but it's also very, very expensive," writes Minter in *Junkyard Planet*. That's the best case. In many developing countries, Minter explains, gold is "removed using highly corrosive acids, often without the benefit of safety equipment for the workers. Once the acids are used up, they're often dumped in rivers and other open bodies of water."

Those methods are major health and environmental hazards, but they have a crucial advantage: they're cheap and easy. Getting copper by burning wire doesn't require special equipment or heavy machinery, just gasoline and matches. That's why, all over the developing world, from Mumbai to Mexico City, laborers risk their lives burning copper wire and dousing circuit boards with chemicals to retrieve the metals that someone in the DRC or Colombia risked *their* life to dig up in the first place.

The damage from these kinds of operations is well documented. Several studies of Guiyu, home to China's biggest e-waste recycling complex, have found extremely high levels of lead and other toxins in the blood of children living nearby. A 2019 study by Toxics Link, an environmental organization in India, found more than a dozen unlicensed e-waste recycling "hotspots" around Delhi employing some fifty thousand people. In those areas, unprotected workers were exposed to chemical vapors, metallic dusts, and acidic effluents, while hazardous wastes, including chemicals and metal dust, were dumped in the open, often close to drains, potentially contaminating the soil and ground water.

Of all the e-waste in the world, lithium batteries are perhaps the most important potential sources for critical metals. After all, where better to get the materials you need to make batteries than from other batteries?

Researchers from the University of Technology Sydney, commissioned by Earthworks, an environmental group, found that full-scale recycling could reduce the need for new supplies of some battery metals by as much as half. Nonetheless, only about 5 percent of all lithium-ion batteries are currently recycled. That's because batteries are probably the most difficult type of e-waste to handle, and certainly the most dangerous.

Though countries like Nigeria are awash in old lithium-ion batteries, there is nowhere on the continent capable of recycling them. Like circuit boards, they need to be exported to facilities overseas. But shipping companies are extremely reluctant to transport these batteries, with good reason: They have a disturbing tendency to burst into flames. If they're punctured, crushed, or overheated, lithium batteries can short-circuit and catch on fire or even explode. Battery fires can reach temperatures topping 1,000 degrees Fahrenheit, and they emit toxic gases. Worse, they can't be extinguished by water or normal firefighting chemicals.

That's why you're not supposed to throw dead lithium batteries out with your regular household trash but, rather, take them to a recycler or hazardous-waste site. No one knows how many people bother to do that, but there are clearly a whole lot who don't. Each year, batteries in everything from old Priuses to sex toys spark hundreds of fires in American scrapyards, landfills, and even on garbage trucks, causing millions of dollars in damage. The numbers are similar in the United Kingdom, Canada, and other countries. Sometimes, badly made or mishandled batteries explode even before they're junked. Fires sparked by faulty e-bikes and scooters killed more than a dozen people in New York City in 2023 alone, and killed and injured many more in other cities across the country.

So, what's an e-waste aggregator in a place like Nigeria to do with the batteries from all those phones and other devices? Often, the batteries are simply dumped in local landfills, where they can leak toxic chemicals—and, of course, catch on fire. Some unscrupulous exporters just mislabel them as something else, handing out a few bribes to make sure officials don't look too closely inside the shipping containers. The results are predictable: Improperly stored lithium batteries that were listed as "spare parts and accessories" ignited on a Chinese container ship in 2020. "I've heard there's a major fire every six months," says Eric Frederickson, vice president of operations at Call2Recycle, America's largest battery-collection organization, "but you never hear about most of them, because they just tip the container over the side of the boat."

Reinhardt Smit is trying a different approach. Smit is the supply chain director of Closing the Loop, a Netherlands-based startup that aims to recycle phones from Africa, but with certifiably sound environmental and social methods. No burned cables, no trashed plastics, no unprotected workers; every step of the process handled responsibly and safely, just the way Western consumers like it.

In a 2021 pilot project, Closing the Loop collected and shipped five tons of phones—plastic, batteries, cables, and all—from Nigeria to a recycler in Belgium, in what the company claims was the first-ever legally sanctioned shipment of such. From a sustainability standpoint, it was a success. From a business perspective, however, it was a money-loser. Clean recycling, it turns out, is hideously expensive.

There are extra costs all throughout the process. To gather the phones, Closing the Loop partnered with Hinckley Recycling, a Lagos-based operation that is one of Nigeria's two—yes, there are only two—fully licensed e-waste recyclers. At the Hinckley compound on the outskirts of Lagos, workers dismantle old phones, computers, and televisions in a clean and well-lit warehouse while wearing reflective safety vests and protective gloves. Without a doubt, this is a safer and more humane setup than squatting in a dumpsite and smashing phones with a hammer. But protecting their workers and complying with other government regulations drives up Hinckley's costs.

Another significant problem is that some components, like plastic casings, have no buyers. "Everything can be recycled, in theory, but it costs money. If you run a smelter, you need to be able to recover your costs," a spokesperson from Umicore says. This points up a uniquely confounding aspect of the recycling industry: the question of who pays whom. It's easy to find someone who will pay to take some materials, like copper, off your hands. But for other materials, like low-grade plastic, you're the one who has to pay someone else to dispose of them. It all hinges on how much money there is at the end of the reverse supply chain. There's usually some, but not always enough to cover the costs of gathering and processing the materials. "If I recycle every component in a phone, I lose money," sums up Adrian Clewes, managing director at Hinckley Recycling. Most of the material in a phone is plastic. Almost no one wants junk plastic. So, Closing the Loop has to pay a recycler to accept that part of its shipment.

Clewes talks about "positive" and "negative" fractions, meaning components and materials Hinckley can make money from, and those that cost money to dispose of. Many fractions slide back and forth between positive and negative territory depending on the prices for raw materials.

Imagine you're an operator like Abubakar. You send out a bunch of guys to gather up phones, smash them open, and get the circuit boards. You end up with a pile of boards containing, say, a total of one pound of copper. Let's say that it would cost a smelter $2 to extract that copper. If the market price of pure copper is $3 a pound, the smelter will buy your circuit boards, because they can recycle them for a profit. But if the price of copper drops down to $1, the smelter will stop buying your boards. Then you've got a pile of printed circuit boards taking up space in your warehouse. Now they're just trash. But that trash can turn back into the proverbial treasure if prices shift again. If you have the space to spare, you can wait for prices to go back up. If you don't, you might be tempted to just dump those circuit boards somewhere.

Then there's the problem of batteries. To get a shipping company to take their batteries from Nigeria to a recycler in Europe, Closing the Loop came up with a clumsy workaround: They put the batteries inside barrels filled with protective sand. That makes things safe, but it also means the company has to pay to transport hundreds of pounds of worthless sand with every shipment.

On top of that are administrative costs. Ironically, international regulations designed to keep rich countries from dumping hazardous waste in poorer ones are now an obstacle to getting hazardous waste *out* of those poorer countries. Foremost among these rules is the Basel Convention, an international agreement that requires any ship carrying e-waste to receive approval from the exporting and importing countries, *and* consent from any additional countries where that ship might stop en route. A ship sailing from Nigeria to Belgium or China is likely to stop in several ports along the way. That creates a mind-boggling amount of bureaucratic hassle. "Observing the Basel notifications can be painful. It takes months," says Batteiger. "The Basel Convention is valuable—without it there would be more dumping—but it has the side effect of blocking exports from the developing world to industrialized countries."

Taken together, all the extra costs involved in legally, thoroughly, and responsibly recycling phones and other e-waste make it almost impossible to turn a profit off the value of the recovered metals alone. Someone has to subsidize those extra costs. Smit's idea is to convince green-minded corporations to make up the difference by paying Closing the Loop to recycle one dead African phone for every new one those corporations buy. The concept is a lot like selling carbon offsets, but for e-waste. It's getting some traction. Closing the Loop now operates in about ten African countries and has collected several million dead electronic devices. Its goal is to collect two million phones every year in the near future. That sounds like a lot, but, as the company's founder, Joost de Kluijver, admits, it's a drop in the bucket. "There are two *billion* phones sold every year," he says. "We can't collect all that."

Meanwhile, the informal sector in Nigeria is generating exponentially more jobs for desperately poor people. It's hard to say which model is "better." That question begs another: Better for whom? Unregulated plastic dumping, wire burning, and working without safety equipment definitely don't measure up to Western environmental and labor standards. But those standards aren't top-of-mind in places like Nigeria, where most people can barely afford to feed and house themselves, let alone educate their children. As with the artisanal cobalt miners in the Democratic Republic of the Congo, the problem isn't so much the actual work e-waste scrappers are carrying out, but the conditions in which they're doing so.

The world's top locus of lithium-battery recycling is, no surprise, China. As of 2021, China was home to 80 percent of global battery-recycling capacity. Much of that is provided by a subsidiary of CATL, the world's biggest battery maker. With locations across the country, CATL has the capacity to recycle 120,000 tons of batteries each year, and it is investing billions in new plants.

But the industry is growing, and its center of gravity is shifting. A towering wave of investment is pouring into e-waste recycling in the United States and elsewhere, especially targeting electric-vehicle batteries. Umicore started as a mining outfit more than two hundred years ago but repositioned to emphasize recycling around the beginning of this century. It opened its first pilot battery-recycling plant in 2011, specifically to be ready for the expected wave of electric-vehicle batteries. Major automakers are partnering with recyclers or even launching their own recycling plants, recognizing that used batteries could provide a cheaper, cleaner, and much more publicly appealing way to feed their ever-growing hunger for critical metals. "It is clear that the biggest mine of the future has to be the car that we already built," said Mercedes-Benz Group chairman Ola Källenius at a 2021 global climate summit.

Leading the charge in the United States is Redwood Materials. Redwood has built, in a remote part of Nevada, an enormous lithium-ion car-battery recycling operation. The company has deals with Tesla, Amazon, and Volkswagen, and has attracted nearly $2 billion in investment capital.

One of Redwood's closest competitors in North America is Canada-based Li-Cycle. Ajay Kochhar, a chemical engineer with neatly combed-back black hair, cofounded the company in 2016 with a friend he'd worked with at a metallurgy firm. "We heard lots of people say, 'You guys are too early. Electric vehicles aren't a thing yet,'" he tells me with a big smile at the company's headquarters in Kingston, Ontario. That year, the company produced its first batch of shredded battery material. "It took us three months to get twenty tons," says Kochhar. Li-Cycle went public in 2021, valued at nearly $1.7 billion. By early 2023, it employed more than four hundred people, had opened facilities in Arizona, New York, and other places with a combined capacity of fifty-one thousand tons of battery material per year, and had deals with General Motors and Glencore.

Li-Cycle's original Kingston operation is relatively small, but it encompasses much of the battery recycling process. On the day of my visit, a truckload of consumer batteries from laptops, cell phones, and power tools has come in, delivered by a "recycling" company that picked them up from a big hardware store. It all gets loaded onto a conveyor belt, where workers strip off whatever can easily be removed—plastic casing, foam packaging, and so on—and check the labels on each one to make sure they are lithium-ion batteries and not some other type. It's a surprisingly labor-intensive process. A human being has to pick up every single battery. Electric car batteries, which are made up of hundreds of small cells packaged in a housing that can be as big as a mattress, also have to be taken apart by hand.

Once it has passed inspection, the mismatched flock of batteries continues up the conveyor belt until it gets dumped into a column of water that carries it down into a shredder. The machine's mighty steel teeth rip the batteries into tiny pieces, like tree branches run through a wood chipper. Whatever plastic is still mixed in floats to the top of the water and is skimmed off.

The metals are separated out in a series of further steps. Breakfast-cereal-sized flakes of copper and aluminum get poured into large, heavy plastic bags. Most of what's left is what's known as black mass—a grainy sludge of other battery ingredients, including lithium, cobalt, and nickel. Li-Cycle will sell the copper and aluminum flakes to a company like Glencore, which will handle the final stage of melting them down and reconstituting them as pure metal, completing the recycling cycle. Similarly, Li-Cycle will sell the black mass to some other company that will use chemical processes to separate out the metals for reuse. (Li-Cycle aims to handle more of these steps in their future plants.) Recycling everything, however, is tough. A certain amount of some metals—especially lithium—is usually left behind.

Ironically, one of the major difficulties facing battery recyclers like Li-Cycle at the moment is a shortage of batteries to recycle. The industry is building facilities at a breakneck pace, so fast that analysts expect factory capacity to outstrip available feedstock in the coming years. (In fact, at the time of this writing in early 2024, Li-Cycle's rapid expansion had run it into financial troubles, causing the company to lay off dozens of employees and suspend work on a plant in Rochester, New York.) At this point, most battery recyclers are relying heavily on pre-consumer factory scrap and defective batteries sent to them from manufacturers. Electric cars themselves are still so new that, so far, very few have been junked. (And there are hardly any at all in poorer countries like Nigeria.) Even when drivers do dispose of their EVs, the batteries are often snapped up for use in things like off-grid power storage—an idea we'll come back to. Most of the billions of small lithium batteries in consumer goods don't get collected at all.

That leaves companies like Li-Cycle hustling for supplies. "We've looked at doing the collection ourselves, but the economics are very challenging," says Kochhar. "There's no clear solution on how to get these things out of people's drawers."

How, then, can more of the e-waste that's just lying around be brought into the reverse supply chain? One approach is to shift at least some of the onus of recycling off of consumers and onto the companies that manufactured the gadgets in the first place. The concept is known as extended producer responsibility. China and much of Europe have adopted it into laws governing not only e-waste but glass, plastics, and even cars. Sometimes that just means charging manufacturers a small extra fee to help cover the costs of recycling their products. Sometimes it imposes serious obligations. Under European Union rules, for instance, car manufacturers are responsible for collecting and recycling vehicles that have reached the end of their lives. Since 2018, China has required manufacturers to collect and recycle lithium-ion batteries and mandates a minimum amount of recycled materials that must be used to make new batteries. There are many loopholes and shortcomings in these laws, of course, but, on balance, they provide a boost to the recycling ecosystem. According to CATL, China is currently recycling at least half of its batteries. "In North America, it's mainly us and Redwood" for battery recycling, says Kochhar. "There are many more in Europe. The regulations are stronger there, and extended producer responsibility laws mean there [is] more collection." In China, "the supply chain for recycling is very crowded," he says. "Their battery-recycling industry is way ahead of what we're doing here."

Direct government support will probably also be necessary to scale up e-waste and scrap recycling to the levels that will really make a difference. The logic of markets and the logic of sustainability are frequently at cross purposes, and such is typically the case with recycling. As a strictly economic proposition, recycling doesn't always make sense. It's often cheaper to mine fresh metals than to recycle them.

Some products are extremely difficult to recycle—including, ironically, some of the machines most crucial to the energy transition: permanent magnets, solar panels, and wind turbines. According to a 2022 report by a leading European research organization, "there is currently no large-scale commercially available recycling of end-of-life permanent magnets." Less than 5 percent of rare earth magnets are currently recycled. As for solar panels, they contain tiny amounts of valuable materials like copper, silver, and polysilicon, but also toxic chemicals that require expensive treatment. As a result, it costs roughly $20 to $30 to recycle a solar panel, but only $1 to $2 to send it to the dump. So, it's no surprise that most end-of-life panels—as many as nine in ten, according to one expert—wind up in landfills. The giant blades on wind turbines are also tremendously difficult to recycle. More than 720,000 tons of them are likely to be trashed by 2040.

All of that is an argument for government intervention to alter the economic calculus by providing tax breaks or subsidies to recyclers. Why not? China gives tax breaks to its metal recyclers. America spends billions subsidizing the fossil-fuel industry and billions more propping up farmers. Congress has not yet fully grasped the connection, however. Two major spending packages passed in the early 2020s direct some $370 billion toward promoting renewable energy, including nearly $40 billion for nuclear energy, but earmark only a couple billion for recycling. (Curiously, the US government does operate a small electronics-recycling business of its own—as a work program in federal prisons.)

New technologies could also help. As Minter points out in *Junkyard Planet*, American cars were once considered almost impossible to recycle. As a result, the rusting hulks of millions of abandoned autos littered the nation's roadsides, leaking toxic fluids into fields and streams. The invention of relatively small metal shredders, powerful enough to tear a car apart but affordable for scrapyard owners, was a key factor in changing all that. Today, according to Minter, America recycles the metal in nearly all of its cars. "Collectively, [abandoned cars] formed one of the most serious environmental crises in the United States—and then, due to a scrapyard innovation, the problem was solved," he writes.

One particularly acute problem with e-waste is the complexity of the devices. Phones and other electronics are made with hundreds of different substances. As we've seen, the most valuable and easily accessed substances get cherry-picked out, while the rest are often junked. Rare earths, in particular, are extremely difficult to extract, as they are present only in trace amounts.

With that in mind, British researchers are working on portable, inexpensive chemical reactors that could make it much easier for waste pickers to extract rare earths from e-waste. The US government is funding a panoply of other research projects aimed at increasing the efficiency of e-waste recycling efforts. Corporations are trying, too. In Texas, Apple is trialing a robot that can disassemble two hundred iPhones per hour, breaking them apart to make extracting materials easier.

Corporations and researchers are also looking to another form of metal-rich trash: tailings and other mining waste. These byproducts often hold small amounts of other metals besides the ones targeted by a specific mine or smelter. Rio Tinto, for instance, is experimenting with chemical processes to extract lithium held in waste rock at a Californian boron mine. A Canadian startup is working to separate out rare earths found in a Brazilian tin mine's tailings. It's a promising concept, but, so far, no one is making it work on a large scale, partly because of high costs.

There is a labor force, however, that will work all day long for zero pay to extract those metals: plants. Several species of trees, shrubs, and other plants called hyperaccumulators suck up tiny specks of metals through their roots and concentrate them in their sap, stems, or leaves. The sap of *Pycnandra acuminata* trees, for instance, which grow on the nickel-rich Pacific island of New Caledonia, can contain more than 25 percent nickel. Other plants can slurp up cobalt, zinc, lithium, and other metals. Researchers in the UK, Australia, and elsewhere are experimenting with a range of hyperaccumulating grasses, flowers, and other plants, putting them to work sucking metals out of piles of mining waste or polluted soil.

If plant-based metal harvesting, known as phytomining, can be made to work at scale, it could help solve several problems at once. The plants could both reduce toxic waste in the soil and boost supplies of critical metals. Farmers in mining-polluted areas could conceivably cultivate hyperaccumulating plants as a cash crop. A handful of startup companies are working to commercialize the idea.

As always, though, there are potential downsides. In the late 1990s, a company called Viridian Resources convinced a county in northern Oregon to plant fifty acres of a hyperaccumulating plant called yellow-tuft alyssum, which they promised would pull toxic metals out of the area's soil. The plant turned invasive, spreading out wildly and threatening local wildflowers, and Viridian went bankrupt, leaving the cleanup to the local government and volunteers.

Everyone agrees that recycling is going to be an increasingly important way of getting the critical metals we need for the energy transition. Analysts believe that using recovered metals has the potential to reduce demand for mined lithium by 25 percent, for cobalt and nickel by 35 percent, and copper by 55 percent by 2040. An expanded recycling industry could also create thousands, perhaps millions of jobs. But recycling is not a silver-bullet solution to the critical-metal supply conundrum. It's extremely complicated, incurs serious social and environmental costs, and, moreover, is simply inadequate to fill the demand for these metals in the Electro-Digital Age. Unlike the cardboard in boxes or the glass in beer bottles, metal isn't usually something you use once and then recycle. Countless tons of copper, for instance, are locked into buildings and machinery where they will remain for decades.

Even if we recycled 100 percent of the critical metals currently in use around the world, we'd still have to mine more, because demand keeps growing. And we'll never be able to recycle all the metals we use. Some material is always lost in the process, whether it falls off a truck en route to the shredder or is vaporized in a blast furnace. Some is simply ignored by recyclers focusing on the most valuable elements in a given product, such as the cobalt and nickel in batteries. The lithium in those batteries is typically an afterthought; at the moment, the world recycles less than 1 percent of the lithium it uses. Some metals, like rare earths, are so difficult to recover at scale that we are likely to depend on mining fresh supplies for decades to come.

"Nothing—nothing—is 100 percent recyclable, and many things, including things we think are recyclable, like iPhone touch screens, are unrecyclable," writes Minter. "Everyone from the local junkyard to Apple to the US government would be doing the planet a very big favor if they stopped implying otherwise, and instead conveyed a more realistic picture of what recycling can and can't do."

In short: Recycling will help provide *some* of the critical materials we need. From the industrialized world to emerging economies, we should be recycling more metal. We should be doing it better. We have to figure out how to collect and process as much metal as possible while making sure the people actually doing the work are reasonably protected and fairly paid. And all of that should happen with the lowest carbon emissions possible.

But recycling alone will never replace mining. That means we—governments, advocacy groups, consumers—need to think beyond the question of "How can we increase the supply of critical metals?" We need to think more deeply about the demand implied by that question. We need to consider how we can reduce our hunger for critical metals in the first place.

PART THREE

BETTER THAN RECYCLING

As we peer into society's future, we—you and I, and our government—must avoid the impulse to live only for today, plundering, for our own ease and convenience, the precious resources of tomorrow.

—DWIGHT D. EISENHOWER, US PRESIDENT (1953–61)

CHAPTER 10

New Lives for Old Things

t didn't seem like a life-altering event when Kyle Wiens accidentally dropped his Apple iBook off his dorm room bed one day in 2003. The laptop landed on its corner, where the plug connects to the machine, and when Wiens picked it up, the little light indicating that electricity was flowing into the computer was dark. Wiens guessed the problem was a loose wire or broken soldering joint, something that should be pretty easy to fix. He was a handy guy, an engineering student at California Polytechnic State University in San Luis Obispo. His grandfather, who had attended the same school years earlier, had given him a soldering iron as a present when he left for college for just such small repairs.

Wiens tried to open up the machine and take it apart. He was confounded almost immediately—there were all kinds of small tabs and latches, and the whole apparatus was bafflingly complicated. Wiens had worked in an Apple Store in high school, so he knew there were repair manuals that could walk him through what needed doing. He tried hunting one down on Google, but, to his astonishment, couldn't find one.

So, he just winged it. With the help of a friend, Luke Soules, he took the machine to pieces on his dorm room floor. It took them two days, and they lost a few screws and broke a few latches along the way, but they finally managed to find the broken connection and mend it with a drop of fresh solder. They put the computer back together, and it worked fine.

"Why was that so hard?" Wiens wondered. Why couldn't he find anything online about how to repair his laptop? He did more research and learned that the problem wasn't that nobody had made Apple repair manuals available on the internet. The problem was that websites that had posted such manuals had been ordered to take them down by Apple. In other words, one of the world's leading consumer-tech companies was actively working to make it harder for consumers to fix its products.

"That really pissed me off," Wiens says. "The idea that this information was forbidden fruit did not sit well." He'd paid good money to buy his computer from Apple. It wasn't theirs anymore. It was his. Why shouldn't he be able to fix it?

It was time, Wiens decided, to fight for his rights. Specifically, his right to repair.

Sooner or later, all electronic products will stop working. Some part will get fried. Some internal system will slip out of whack. Entropy will inevitably collect its due. What do you do then? If the product in question is a car, no one would answer, "Recycle it and get a new one." Recycling that car would mean tearing it to pieces, separating out the plastic and glass and metals that it's made of, then smelting those down so they can be sold back into the market. Recycling any electronic product similarly means extracting all the constituent materials of that highly engineered product and reverting them back to their original state—their most basic, literally elemental form. By and large, recycling is the most inefficient, most labor- and energy-intensive method of getting further use out of just about any given product.

When it comes to a malfunctioning car, most people start by trying to repair it. You can do that yourself or take it to the dealer you bought it from or to an independent repair shop. When it comes to a whole range of smaller products, however, everything from lamps to vacuums, it's not nearly as common or easy to fix things when they break down. Especially when it comes to digital devices—like phones, tablets, and laptops—the default is: When it breaks, just replace it.

This is no accident. It's the result of a deliberate corporate strategy. It was that strategy that Wiens set out to battle.

The most obvious reason to fix an electronic product (or any product, really) rather than replace it is simple: Repair is usually cheaper. To that personal motive, add a planetary one: the more electronic products whose lives we extend via repair, the fewer natural resources and the less energy we need to manufacture new electronic products. The imperative to fix things, writes journalist Aaron Perzanowski in *The Right to Repair*, "grows out of a recognition that resources are finite, that the planet is small, and that a culture that overlooks those facts imperils its future. Repair allows us to extract maximum value from the artifacts we create."

Each broken cell phone or hair dryer contains only small amounts of copper, nickel, and other metals, but keeping millions or billions of them in use obviates the need for huge amounts of mining. Remember that it took something like seventy-five pounds of raw materials to manufacture your phone, generating heaps of waste in the process. Digging all that material up burns a lot of energy, generating greenhouse gases as well. In his book *Electrify: An Optimist's Playbook for Our Clean Energy Future*, engineer and MacArthur "Genius Grant" recipient Saul Griffith analyzed the energy used and carbon emitted per kilogram of a range of products. His conclusion: "To reduce environmental impact, you can lower the weight of a thing, or you can use a different material altogether. But making the thing last longer is key." Extending the lifetime of all the smartphones in the European Union by just one year would prevent the release of 2.1 million metric tons of carbon dioxide per year by 2030, according to a study by a consortium of European environmental groups. That's the equivalent of taking more than a million cars off the road.

Wiens and Soules launched their crusade by writing their own laptop-repair manuals and posting them online. The response from consumers was enthusiastic. "We got like thirty thousand hits the first day," Wiens says. They branched out into other products, and soon a dorm room business, dubbed iFixit, was born. By the time he graduated in 2005, Wiens had decided to try making a career of it.

Today, from its base in San Luis Obispo, iFixit hosts a free online repository of more than 103,000 do-it-yourself repair manuals for some fifty-four thousand separate products, from the iPhone 15 Pro to Black+ Decker's handheld pet-hair vacuum. Some were written by staff members, many by volunteers. Millions of people visit the website every month. The company makes money selling tools and replacement parts and from consulting services. Wiens declined to share his revenue numbers, but there's enough cash coming in for him to employ about 170 people.

"Our mission is to teach everyone to fix everything," he says. "We want to make it easy to share knowledge and simple to do repairs." To do that at scale, however, Wiens decided it wasn't enough to rely on individuals figuring out solutions and sharing what they learned. He wanted to force the corporations that make that gear to help. Or, at least, to stop making it harder.

What's good for consumers and the planet is not always good for corporations, of course. Apple makes money if you buy a new iPhone; it makes nothing if you get your old iPhone fixed at an independent repair shop. That's why companies of all sorts have, for decades, discouraged repair and encouraged customers to instead buy the newest, latest thing. As far back as the 1920s, Ford and General Motors began introducing new car models each year with the explicit intention of getting drivers to trade in their functional but no-longer-fashionable older flivvers. Planned obsolescence, as it became known, was celebrated as a tool to keep the consumer economy humming.

America's burgeoning wealth helped undermine the practice of repairing things. "In the nineteenth century, and well into the twentieth," writes Adam Minter in *Secondhand*, "thrift wasn't a matter of choice or virtue. It was a necessity. The essentials of daily life—clothes, kitchenware, tools, furniture—were expensive and intended to last for years if not lifetimes. Repair was a way to ensure that they did." But as Americans got richer and manufacturing got cheaper, fixing increasingly fell out of fashion. In 1966, according to Perzanowski, some two hundred thousand Americans worked as home appliance repairers; by 2023, that number had dropped to about forty thousand. The world of digital gear is on the same trajectory. The number of electronic and computer repair shops in the United States dropped from 59,200 in 2013 to 45,830 in 2023.

Ironically, it's often much easier to get electronics fixed in developing countries than in rich ones. In places like Nigeria, where many people subsist on just a few dollars a day, "the essentials of daily life" are still expensive and worth repairing. The Ikeja electronics market, where I met Baba Anwar, teems with tiny kiosks and elbow-to-elbow stalls where tradesmen will replace a broken phone screen or swap a new hard drive into your laptop while you wait. They're not certified by any big corporations, but their skills are top-notch. Lagos is particularly renowned for the quality of its electronics repair workers, but you can find people doing the same work in cities from Delhi to Cairo—everywhere people don't have the wasteful luxury of constantly buying new electronics.

This is not entirely the fault of affluent consumers. The electronics industry deliberately makes their products difficult to repair. Take a look at your cell phone, tablet, or laptop. Chances are, there is no easy way to even open it up to see its components the way you can pop open a car's hood, let alone swap in a new battery the way you can with a flashlight. That's by design. Frequently broken parts, like cell phone screens, are held in place with glue, making them hard to remove without damaging the phone. Other parts are fastened with unusually shaped screws that require special screwdrivers. Manufacturers deliberately discourage DIY repairs with claims that they are unsafe or will void the product's warranty.

"The 2019 iMac manual cautions that 'Your iMac doesn't have any user-serviceable parts, except for the memory . . . Disassembling your iMac may damage it or may cause injury to you,'" writes Perzanowski. "Apple offers similar warnings to iPhone owners, who are told in no uncertain terms 'Don't open iPhone and don't attempt to repair iPhone yourself.'" Local fix-it shops are also hamstrung. Parts, manuals, and necessary software are often made available only to high-priced authorized service providers.

By the early 2010s, Wiens had decided that the only way electronics companies would change their ways was if they were forced to. So, he shifted from publishing manuals to political advocacy. Ever since, Wiens and other activists have been lobbying local, state, and federal legislators to enact right-to-repair laws that would obligate manufacturers to make manuals, tools, and parts available to everyone. Similar rules already cover automakers; that's why you can take your balky Volkswagen either to an official dealer *or* an independent local shop. Well over one hundred right-to-repair bills were introduced in state legislatures in the 2010s. Every single one of them was shot down.

It wasn't exactly a fair fight. On one side were Wiens, local activists, and consumer-advocacy groups; on the other were Apple, Microsoft, Amazon, Google, Tesla, T-Mobile, and other giant corporations, all of which have lobbied against right-to-repair bills. "The industry was able to sow a lot of FUD—that's fear, uncertainty and doubt," says Wiens, who has testified in support of dozens of these bills. "I have a lot of sympathy for these state legislators, because they have so many complex issues coming across them all the time, and then they have Apple come in and say, 'If you do this thing, you'd be the first state ever to do it, and that will disrupt all commerce in America.' You don't want to be the person who screwed up the technology industry in the US."

Manufacturers argued that their restrictions were necessary to safeguard trade secrets and to protect people from hurting themselves by messing with their machines. "Apple told Nebraska lawmakers that the bill would turn the state into a 'Mecca for bad actors,' predicting that hackers and other nefarious figures would flock to the state to exploit consumers," writes Perzanowski. "And in California, it warned that consumers were at risk of physical injury if they attempted to swap out their iPhone batteries. Wahl cautioned that repair of its hair clippers could cause fires, while Dyson and LG issued unfounded warnings that the right to repair could put consumers' personal safety at risk by allowing repair personnel in their homes who had not cleared background checks."

Apple CEO Tim Cook all but admitted the real reason behind his company's opposition to these laws in a 2019 letter to investors. The previous year, it had come to light that Apple had been deliberately slowing down the performance of some older iPhones. Customers were furious. In a bid to placate them, Apple temporarily reduced the price of an authorized battery replacement from $79 to $29. Some eleven million iPhone owners took the deal, giving their old phones a new lease on life. Result: Sales of new phones dropped. That was a big problem for the company, since iPhone sales provide the bulk of Apple's revenue. "iPhone upgrades . . . were not as strong as we thought they would be," Cook wrote, blaming, in part, "customers taking advantage of significantly reduced pricing for iPhone battery replacements." The cut-rate replacement program was soon scrapped. As of 2023, Apple was charging $135 to replace an iPhone's battery.

But years of pressure from Wiens and others finally helped achieve a breakthrough. In 2021, the Federal Trade Commission published a landmark report that scrutinized all of the tech industry's objections to repair laws. Its authors concluded that device makers' concerns could either be addressed with some modifications or simply had no merit in the first place. "There is scant evidence to support manufacturers' justifications for repair restrictions," the report concludes tartly. "Although manufacturers have offered numerous explanations for their repair restrictions, the majority are not supported by the record." President Biden soon followed up with an executive order that encouraged the FTC "to limit powerful equipment manufacturers from restricting people's ability to use independent repair shops or do DIY repairs." The following year, the FTC put some bite into its bark, fining Harley-Davidson and Westinghouse for illegally restricting customers' ability to repair their products.

By then, the European Union had already adopted rules designed to make it easier to repair home appliances, including requiring manufacturers to make spare parts available and ensure that they can be replaced with common tools. The UK also rolled out regulations obliging electrical-appliance manufacturers to make spare parts available to consumers.

Manufacturers have started to get the message. Since 2021, Motorola, Samsung, and others have made a complete 180, allowing independent repair shops access to parts and tools and even partnering with iFixit to help customers repair their products. "They've shifted from being antagonists to business partners," says Wiens. "They are not coming of their own free will," he is quick to add. "It took getting to the point where it was clear that legislation was inevitable for the companies to come on board."

Not that all of them did so with the best of grace. In 2022, Apple rolled out a self-service repair program. It posted online manuals for some of its products and offered to send users the same set of tools used in the company's repair facilities. That sounded good. But when a *New York Times* reporter tried out the service, he found that "it involved first placing a $1,210 hold on my credit card to rent 75 pounds of repair equipment, which arrived at my door in hard plastic cases. The process was then so unforgiving that I destroyed my iPhone screen in a split second with an irreversible error." Critics called it "malicious compliance," an effort designed to fail.

That same year, however, the American repair movement notched its first major legislative victory. New York State enacted a right-to-repair law, which iFixit helped write, requiring manufacturers to make service information, parts, and tools available to the public. Lobbyists from Microsoft, Apple, and other tech companies managed to punch some big loopholes in it—the law covers only digital electronics, leaving out home appliances and many other devices, and applies only to machines manufactured after July 1, 2023. Still, says Wiens, "we were thrilled. It was a big moment." Momentum is growing. Several months later, Minnesota adopted a similar law. In October of 2023, California also passed one—with the surprise support of Apple. Someday soon, all Americans may have the freedom to fix.

A FOUR-HOUR DRIVE SOUTHEAST FROM IFIXIT'S HEADQUARTERS, IN THE ANTE-lope Valley, an arid expanse on the edge of the Mojave Desert, a different kind of effort is underway to extend the lives of electronic gear. For years, the valley's farms have been giving way to spreading suburbs, as is the case in much of inland California. But this particular area is being shaped by another economic driver: renewable energy. Enormous solar farms have taken over thousands of acres in the valley. Forests of wind turbines stand on the hills ringing the broad valley floor. Chinese electric-vehicle giant BYD has an electric-bus factory on the outskirts of the town of Lancaster. Less than a mile from there, on a patch of table-flat bare dirt studded with white, shipping-container-sized metal boxes, Freeman Hall is trying to build his own lucrative niche in this electrified landscape.

A keening whine emanates from the containers. It's the sound of power. Each of the dozens of boxes is stuffed with used batteries rescued from defunct electric vehicles—many of them good old Nissan Leafs. Those batteries are a little too worn down to give the performance an EV requires, but they can still hold a decent charge. Packaged together by the hundreds, they can hold a lot of energy. The batteries in all those containers assembled by Hall's startup company, B2U Storage Solutions, collectively hold enough electricity to power 3,400 homes.

Hall, a neatly bearded thirtysomething, launched his company in 2019 with the idea of giving EV batteries a second life. "I used to work in the solar industry, and I learned there is a lot of demand for power in the evenings, when the sun isn't shining," he says. "In 2017, there were already about four hundred thousand Leafs on the road. My partners and I realized that a second life could work for lots of those batteries." They landed deals with Nissan and, later, Honda, in which B2U collects used or defective vehicles, manually extracts the batteries, and installs them in the containers. The idea behind the business is simple: During the day, B2U's adjoining solar farm charges up the batteries. At day's end, B2U sells that power to electric utilities. The sun's energy delivered at night.

Solar and wind power can only be viable at scale if there's a way to stockpile the energy they generate when the sun is shining and the wind blowing for the times when they're not. There has to be a means to bank that power and withdraw it when it's needed. There are several ways to accomplish this. In some places, utilities are building enormous, grid-scale batteries from scratch. The advantage of repurposing used car batteries is that it uses products that already exist. With car batteries, you can build an energy-storage system that doesn't require any new mining, refining, or manufacturing. It's the "reuse" part of the famous "Reduce, Reuse, Recycle" slogan. What's more, used batteries are as much as 70 percent cheaper than new ones for energy-storage systems, according to the consulting firm McKinsey. By 2030, used electric car batteries could store as much as two hundred gigawatt-hours of power per year. That's enough to power almost two million Nissan Leafs.

Of course, reusing an electric-car battery isn't as easy as popping a Duracell out of a flashlight and into a TV remote. The batteries inside Teslas, Bolts, Leafs, and other EVs have no standard format. They come in a wide range of sizes, shapes, and chemistries. It's difficult to get all those different types of batteries to work together in an energy-storage system. Used batteries also have to be tested to make sure there's enough life left in them, a process that is also complicated by all those different designs. No one really knows how long the batteries will last, either. Factors like the temperature they're stored at and the rate at which they are charged and discharged can affect their longevity. Hall thinks B2U will get about five and a half years out of its batteries, but he won't know for sure until they quit working.

Nonetheless, the concept is attracting a lot of interest. South Korean automaker Kia has partnered with Germany's national railway company on a project to reuse car batteries in electricity-storage systems. Nissan is putting old Leaf battery packs to work powering its American headquarters in Tennessee. The auto giant has also tinkered with repurposing Leaf batteries to power streetlights and even 7-Eleven stores in its home country of Japan. General Motors, Toyota, and others are experimenting with similar ideas.

B2U has competition from other startups, too. California-based Smartville has a system that can combine batteries from several different carmakers in a single storage unit. Up near Vancouver, Moment Energy collects used Mercedes batteries from dealers and adapts them to provide off-grid power to remote homes and businesses. At the moment, all of these efforts are early stage and small scale, but Hall expects car-battery reuse to grow fast, in step with the overall electric-vehicle industry. "The market is doubling every two years," says Hall. "We're only in the first inning,"

In yet another part of California, the Sierra Foothills east of Sacramento, a nonprofit group called Good Sun is reusing a different piece of renewable-energy hardware: old solar panels. Like batteries, photovoltaic panels degrade over time. Each year that they spend soaking up sunshine reduces their efficiency. Used panels still produce electricity, but not as much as they once did, and generally much less than the newer models that are always coming out. So, when solar-power users near Good Sun's headquarters in the town of Grass Valley upgrade their gear, the group collects the used panels and redeploys them at local homeless shelters and schools. They've also taken panels to Africa for use at an orphanage and a hospital.

Solar panels are made from their own set of critical minerals with their own associated problems. Chief among them is polysilicon, the main material making up the sunlight-absorbing cells in almost all solar panels. As you'd expect, China dominates production of this material. The plants that crank it out are mostly powered by coal—again, the world's dirtiest fuel source is burned to create the machinery of "clean" energy. To make matters worse, almost half of the world's supply of polysilicon is manufactured in China's Xinjiang region, homeland of the persecuted Uyghur people. Human rights researchers charge that the Chinese government often forces Uyghur people to work in the poly-silicon plants.

Inside a solar panel's cells, a paste made of silver helps convert the sun's energy into electric current. Today, solar panels consume about 10 percent of all the world's mined silver; by 2050 that figure could climb to more than half. Of course, that has side effects. Runoff from silver mines has contaminated rivers in Peru, Bolivia, and elsewhere.

Reusing solar panels makes even more sense when you realize how difficult they are to recycle. Getting the aluminum frames and copper wiring off isn't tough, but separating the photovoltaic cells from the glass panels to which they are bonded is. The panels are also often full of potentially dangerous metals, like lead and cadmium, that require special handling. In the United States, there are only a handful of outfits that even bother recycling solar panels. It's usually much cheaper to just chuck the unwanted panels in a landfill. (In European countries that mandate producers foot the bill for recycling, however, photovoltaic-module recycling rates are as high as 95 percent.)

The International Renewable Energy Agency estimates as much as seventy-eight million tons of solar-panel waste will be generated world-wide by 2050. The US alone will have to handle as much as ten million tons of junked solar modules in the coming years, according to a 2021 report from the Department of Energy. Nonetheless, the report notes, "there is scant evidence of [photovoltaic] equipment reuse and less than 10 percent of end-of-life modules are recycled in the United States."

The developing world, however, is full of people who are happy to put a cheap, used solar panel to work. The ample sun in much of Africa and Asia compensates for used panels' inefficiency—especially since the alternative is often no electricity at all. The World Bank estimates the worldwide off-grid solar market to be worth nearly $3 billion annually. According to the online publication *Solar Power World*, Afghanistan is a top market, followed by Pakistan and several African countries. Much of that off-grid power is supplied by the more than ten million used solar panels already deployed powering homes, well pumps, Wi-Fi connections, and batteries.

People in developing countries from Asia to Latin America are similarly eager to make use of electronics that residents of richer nations have lost interest in. Americans do reuse some electronics. You can buy refurbished laptops on sites like eBay, and charities like PCs for People collect donated computers to give away to nonprofits and low-income people. Some e-waste recyclers extract functioning components, like hard drives, from computers and resell those parts, but that's not the default practice the way it is in the developing world. More than half of all the computers purchased in places like Egypt and Ghana are likely secondhand. Worldwide, the used cell phone market is estimated at $25 billion. The Ikeja market is home to countless tiny businesses that import used electronics and clean them up for resale. In a narrow little concrete room packed floor-to-ceiling with laptops arrayed like books on library shelves, I watched one of these entrepreneurs polish the scratches off a used laptop and attach a fresh HP sticker to its lid. The computer worked fine, looked brand new, and was likely sold for a fraction of its original price.

That kind of repair and reuse generates countless jobs in the developing world. An estimated thirty thousand people work in fixing and refurbishing electronics in Nigeria alone. Those practices could, and should, create more jobs in rich countries, too. "Repair is skilled, labor-intensive work. Unlike manufacturing, it is difficult to automate and tends to benefit local, small businesses rather than global giants," Perzanowski points out.

As with recycling, government subsidies could help grow the repair and reuse industries. They already are in some places. Several states and cities in Austria cover as much as half the cost of fixing devices. France provides individuals with fifty euros to get their bikes fixed. Even in the United States, there are programs to offset the costs of auto repairs for low-income people.

In the short run, there is a clash between the repair and reuse industries, on the one hand, and recycling, on the other. Battery recyclers are already facing a shortage of supply; the more used batteries that get put to work in energy-storage farms like B2U's, the fewer there are in the reverse supply chain for the likes of Li-Cycle.

In the long run, though, everything turns into scrap. A repaired phone will eventually break down irreparably. A reused battery will, sooner or later, degrade so much that it is no longer usable. Repair and reuse only postpone the inevitable demise of any product. But extending the lives of those products slows down the cycle of production, consumption, and disposal, reducing demand for raw materials and energy. Ideally, once those products have truly reached the ends of their lives, they will be put on the reverse supply chain and sent to a scrapyard for recycling.

Repairing, reusing, and ultimately recycling products is much better than throwing them in a garbage dump. But those steps still keep us enmeshed in a cycle of energy- and material-intensive production and consumption. There's only one way to break free of that cycle altogether.

CHAPTER 11

The Road Forward
and How to Travel It

O ne morning in October of 1971, a six-year-old girl named Simone Langenhoff was riding her bike down a road in the southern Netherlands. Suddenly, a careless driver whipped around a blind corner and crashed into her. In an instant, Simone became one of some 450 children killed by automobiles in the Netherlands that year. That horrific toll had been rising steadily as the tiny European nation, like virtually all of its Western peers, came to rely increasingly on cars. In the first part of the twentieth century, the Netherlands had not embraced the automobile with quite the gusto that America had. But, as the country's prosperity grew after the Second World War, so did the number of private cars on its roads. Lethal accidents increased apace.

But Simone's father, Vic Langenhoff, a prominent newspaper columnist, wasn't willing to accept his daughter's death as the price of progress. Instead, he founded an organization to push for safer roads, with the hard-to-ignore slogan of "Stop the Child Murders." The group staged protests, occupied roads, and drew increasing support. Similar activist outfits sprang up in Copenhagen, Montréal, and other places. It took some time, but, gradually, the Dutch activists gained ground. City governments across the Netherlands began to reengineer streets and change policies with the explicit goal of diminishing the use of automobiles on public roads.

The results are dazzling. In today's Amsterdam, it seems *everyone* gets around on bicycle—not just the fit young people you see riding in American cities, but suit-wearing businessmen, wrinkled senior citizens, moms with kids, humans of all shapes, sizes, and cardiovascular abilities. The city itself has been retrofitted to make biking easy, convenient, and, above all, safe. Bike lanes aren't just marked by a stripe of paint; they are physically separated from cars by concrete curbs. And they are knit together to form a continuous, cohesive network that allows riders to travel pretty much all over the city without having to mingle in traffic with automobiles. That bicycle network connects directly to an excellent public transit system. Next to Centraal Station, the city's main train hub, is a huge, multistory bicycle-parking structure crammed with thousands of bicycles. Locals and tourists alike ride to the station, stash their bikes, then take one of the clean and affordable trains to destinations farther afield. More than one out of every three trips taken in Amsterdam is now made by bike. Only one out of every two households even owns a car. In a modern city of nearly one million people, those are astonishing statistics.

What does all that have to do with our need for critical metals? Amsterdam is relevant because it provides a real-world example of the single best way we can reduce the damage done by our consumption of both minerals and energy: by reducing our consumption of *everything*. Above all, of private automobiles.

Humanity's insatiable appetite for energy created the climate-change crisis. The same appetite is now impelling us into an energy transition which carries its own tremendous costs. Switching from fossil fuels to renewables is an improvement, but it's not nearly enough. It's necessary, but not sufficient as a solution. To create a truly sustainable world for the eight billion human beings now living on it, we need to cut down the overall amount of energy we use—preferably without sacrificing the quality of life we enjoy in rich countries, while also raising everyone else up to a similar quality of life.

That's a tall order.

It's also entirely possible.

Take clothing. It's an especially obvious example of a product most of us in the industrialized world could easily consume less of without making our lives any less pleasant. The wastefulness of the garment industry is staggering. Since the beginning of the century, global clothing production has roughly doubled, while the lifespan of each garment has been drastically shortened. With the rise of what is known as fast fashion, mind-boggling numbers of garments are produced each year that are meant to be worn only a few times before getting tossed. "Maybe fashion marketing has convinced you that the industry is now mostly organic and circular, recycling discarded clothes into brand-new ones," writes journalist J. B. MacKinnon in *Sierra* magazine. "In reality, six out of 10 articles of apparel end up in a garbage dump or trash incinerator each year."

On top of all the cotton and other materials that are wasted in the process, junking all these clothes also wastes huge amounts of energy. The mills and factories that produced them, the ships and trucks that transported them, and the stores that sold them all ate up energy. Manufacturing apparel generates as much as 8 percent of global carbon emissions. That's more than most countries. Would your life be much worse if you bought fewer T-shirts? The planet would certainly be better off.

Economists use the concept of embodied energy to measure the total amount of power used to create a given product. Whether that energy came from fossil fuels or renewables, it came at some cost in harm to the environment or human suffering, often both. Everything we use or consume has embodied energy—guitars, sofas, blenders, tires, toothpicks. That goes for "green" products made of natural materials like bamboo, too. It takes lots of energy (and water) to process bamboo into clothing, fabrics, or disposable plates, and more energy is required to ship them. Food also has embodied energy. Broccoli, chicken, rice—it all gets moved around by energy-burning vehicles, stored in energy-burning warehouses, and sold in energy-burning grocery stores. Not to mention the energy used to manufacture the farming equipment, fertilizer, and food-processing machinery needed to grow it and prep it for the market. To reduce your consumption of pretty much anything, then, is to reduce your consumption of energy. Which, in turn, means you are reducing the demand for fossil fuels and/or critical metals.

There are many ways, besides buying fewer clothes, by which we could cut consumption and energy use and barely even notice. Turn down the heating or cooling systems in your home a few degrees. Waste less food. (Worldwide, about a fifth of all food winds up in the trash.) And, of course, repair and reuse digital devices.

By far, however, the most effective single way that we as individuals can make a difference is this: Don't buy a car. Not even an electric one.

That may sound like heresy. But hear me out. I am not saying nobody should own a car. I am not saying that cars are inherently bad. On the contrary, they are tremendously useful. For the millions of people living in suburbs or rural areas where distances are wide and public transport is scant, cars are a necessity. The same goes for people with mobility issues. Plus, millions of people *like* driving and owning cars. They're powerful symbols of independence and status. Under the right circumstances— open road, music cranked, pedal floored—they're lots of fun. Cars are not the problem, in and of themselves. The problem is that there are far too many of them.

Gasoline-powered cars, of course, present the most immediate issue. Currently, America's 280 million–plus cars and passenger trucks are the nation's single largest source of greenhouse gases. Their tailpipe emissions not only warp the climate, they also spew toxins into the air we breathe. A 2013 study by the Massachusetts Institute of Technology found that automobile exhaust contributes to the deaths of fifty-three thousand Americans each year.

Electric vehicles are starting to make a little headway on the emissions front. BloombergNEF estimates that all the EVs on the world's roads in 2022—including electric bikes, cars, and trucks—collectively saved us from burning around 1.5 million barrels of oil per day. That's a lot, but conventionally powered vehicles still burn nearly thirty times that amount. It will take many years before electric vehicles outnumber gasoline-powered ones. Modern cars typically stay on the road for nearly twelve years. Meanwhile, new gasoline and diesel vehicles will continue to be produced. The world's automobile fleet will grow to almost two billion vehicles by 2050, predicts the global research firm IHS Markit, with developing countries spurring most of that growth. Of those two billion vehicles, only about 610 million will be electric. If those predictions bear out, that would mean almost no net change in the number of fossil-fuel-powered vehicles.

Even if we did manufacture two billion electric vehicles to replace all the gas- and diesel-powered ones, we would end up swapping one set of problems for another. The sheer quantity of metals we would need, and the damage we'd cause producing them, is just overwhelming. Electric vehicle and battery production is projected to account for at least half of the additional mineral demand spurred by the energy transition. Reducing demand for cars will do more than anything to reduce demand for critical metals.

There are other reasons to thin the herd of motor vehicles. No matter what powers them, cars inflict a whole range of harm on all of us. We're so accustomed to these costs that most of us barely think about them. We should, though, because their effects are staggering.

Start with the worst harm of all: death. The most dangerous thing the average American does each day is drive somewhere. In 2022, the most recent year for which data is available, an estimated 42,795 people died in motor vehicle crashes in the United States. That means that, every single day, 117 Americans lose their lives in a sudden, violent collision. Nineteen of those people aren't even in a car; they are pedestrians heading somewhere on foot when a motor vehicle plows into them. Worldwide, the carnage is overwhelming: nearly 1.3 million people die and as many as 50 million more are injured in car crashes each year. Since the first such fatality, on a London street in 1896, more than 50 million people have been killed by automobiles.

One reason the death toll keeps climbing is that more Americans than ever are driving large, heavy SUVs and pickup trucks, which are more likely to kill those they hit. All else remaining equal, there's no reason to think a switch to electric vehicles will bring down the accident-casualty totals. Automakers are happily rolling out all kinds of electric pickups and SUVs. Those may be even more dangerous. Thanks to their batteries, electric vehicles tend to be heavier than their gasoline-powered counterparts.

Electric vehicles also cause air pollution. Though they have no tail-pipe emissions, they do have tires, and it turns out that tires have their own emissions. As tires roll over pavement, tiny particles of chemical-infused synthetic rubber break off and float away. In the United States and United Kingdom combined, researchers estimate, some three hundred thousand tons of rubber are released into the environment every year, polluting air, soil, and water. Low-income and nonwhite communities, which tend to be closer to major highways, bear the brunt of such harms.

Then there are the colossal tracts of sheer space, the oceans of real estate—a precious commodity in cities especially—that cars have taken from us. Most automobiles are parked 95 percent of the time, just sitting, immobile and unused, doing nothing except hogging up space. From an economic perspective, this is absurd. The financial firm Morgan Stanley has called cars "the world's most underutilized asset." From a land-use perspective, it's infuriating. If you choose to sacrifice square footage in your private home for a garage or driveway, that's your right. But cars mostly occupy public space. They are parked by the millions on public streets. There is usually no charge for this enormous gift of real estate we collectively bestow on cars. At any given moment, most of that real estate sits empty, waiting for a car to lumber in from some other parking space. The acreage lost is stupendous. Across the United States, there are as many as two billion parking spaces—seven for each motor vehicle. Los Angeles County alone gives up about two hundred square miles of land to parking. Even the supposedly environmentally conscious San Francisco Bay Area contains twice as many parking spaces as it does people. Lined up end-to-end, those spots would circle the Earth—twice. Imagine what else could be done with all that space if it wasn't covered with asphalt. Think of all the parks, sidewalk cafes, playgrounds, and new housing we could have if we reclaimed even half of that acreage from cars.

Cars have so thoroughly dominated our roads for so long that most of us have come to accept their costs and demands without giving them much thought. The thousands of deaths, the massive pollution, the loss of space—they all just seem like facts of life, regrettable but immutable. After all, in the modern world, we *need* cars, right?

Right.

But we don't need nearly so many of them.

Why does anyone own a car? Some people enjoy driving. Some like collecting vehicles. But for most people, cars are a means to an end. They are convenient machines that are always on call to take us where we want to go. The end goal is transportation, not car ownership for its own sake.

America of course, is the car-happiest nation on the planet. As of 2018, there were 867 cars for every thousand people in the US, compared to 520 per thousand in the European Union, 160 in China, and 37 in India. People in those countries somehow get to their jobs, schools, soccer fields, restaurants, and grandparents' houses. So, clearly it's possible for large numbers of people to travel from place to place without quite so many cars.

Seen from that angle, the real issue isn't how to get more metals into the global supply chain to build more cars, it's how to get people to where they want to go with *fewer* cars. Metal is a means to a means: the car. But the end we seek is to be able to get from home to work or school and back safely, quickly, and reliably. There are other ways to do that—more efficient, more healthful, more sustainable ways—than driving a privately owned car. Primary among them is a privately owned machine that predates the car.

When it first hit the streets in the 1880s, the modern bicycle was a revolutionary innovation, the iPhone of its time. In those days, before the automobile was invented, when horses, trains, and feet were the main ways to get around, people were amazed and thrilled by these marvels of science and engineering, these small, light, "silent steeds of steel" astride which they could zip around at astonishing speeds.

The machines were cheaper and cleaner than horses, and they didn't need to be fed, stabled, or coaxed along. They just needed a rider with legs and a bit of nerve. "The exhilaration of bicycling must be felt to be appreciated," enthused the *San Francisco Chronicle* in 1879. "With the wind singing in your ears, and the mind as well as body in a higher plane, there is an ecstasy of triumph over inertia."

The public was smitten. In 1885, a handful of American factories cranked out about eleven thousand bicycles. A little more than a decade later, that number had shot to more than two million. Bicycle racing became the most popular athletic competition in America, drawing bigger crowds than baseball games. (It also produced America's first Black sports star: Marshall "Major" Taylor, a Brooklyn cyclist who became a world champion.)

At the peak of the bicycle's popularity, in March of 1896, an inauspicious historic event was recorded: America's very first traffic accident involving a car. A man named Henry Wells was motoring down a New York City street in one of the first commercial automobiles when he struck a bicyclist, breaking the rider's leg. At the time, no one recognized the incident as the metaphor it was.

Bicycles were about to taken down by another revolutionary means of transportation, an even faster, more advanced, more exciting personal-transport machine: the automobile. The race was won quickly. American bike sales plunged 90 percent between 1899 and 1909, while car sales roared upwards. It took only a few years for the bike to be shunted into a distant second place among forms of personal transport. For decades thereafter, bicycles were largely relegated to the status of toys for children and exercise equipment for adults.

Today, however, we are in the early stages of a second bicycle revolution. Bikes are surging back to the forefront of urban transportation, increasingly recognized as critical tools to fight climate change and reclaim our cities. These elegantly simple, two-wheeled machines are once again changing how millions of people travel from place to place, and the shapes of cities themselves.

The most obvious benefit to traveling by bicycle is that it doesn't emit an ounce of carbon. If even a fraction of the motor vehicle trips taken each day in the United States were instead taken by bike, we could take a serious bite out of carbon dioxide emissions. (In a weird coincidence, the very first bicycle was invented as a response to climate change. A massive volcanic eruption in 1815 lowered global temperatures and ruined crop harvests across Europe. Countless people, and the horses they rode, starved to death. In an effort to provide replacement transportation, a German inventor came up with the first protobicycle.)

Riding is also good for your health. Many studies have shown that regular cycling reduces a person's risk of obesity, diabetes, stroke, heart disease, and other ailments. Plus, bikes are quiet and small. Even if one hits you or you fall off one, you are unlikely to be killed. Fewer cars and more bikes on city streets makes for a safer, more pleasant, human-scale environment.

To American ears, turning millions of drivers into bikers sounds like a pipe dream. But here's the thing: It's already happening. Since 1990, the share of urban trips made by bicycle has grown exponentially in dozens of cities around the world. In addition to bike meccas like

Amsterdam and Copenhagen, major European cities—including Berlin, Paris, Vienna, and Barcelona—have invested heavily in bicycle infrastructure and have seen bike usage double, triple, and quadruple. In Taipei, Shanghai, and other Asian cities, millions of people regularly get to work and school by bike. In Tokyo, the world's biggest megalopolis, 15 percent of all trips are taken by bike.

Even in the car-besotted United States, transportation habits are changing, at least in cities that are working to become more bike-friendly. Since Portland, Oregon, began aggressively expanding bike lanes in the 1990s, the number of residents who cycle to work has grown sixfold. Minneapolis, San Francisco, and Washington, DC, have seen similar surges. Even Los Angeles, ground zero of America's car culture, is building new bike paths and expanding its bike-share program across the city. Those efforts are paying off: Bicycle commuting in LA has doubled since 2005.

Another indicator is the growth of the global bike industry. As recently as 1965, the world produced roughly the same number of cars as it did bikes—around twenty million of each per year. Today, well over one hundred million bikes are manufactured each year, far outstripping car production. Bicycles are now a nearly $70 billion global industry.

You don't even need to own a bike to ride one anymore. In cities all over the world, bike-share programs are sprouting like dandelions. Most common are systems in which the bikes are taken from and returned to docking stations, but there are also app-based, dockless systems which let you find bikes (and scooters) scattered around city streets, rent them for five minutes or five hours, and then just leave them wherever you want. First launched on a large scale in Paris in 2007, bike-share systems are now running in well over one thousand municipalities. The number of share bikes worldwide has shot from fifteen thousand in 2007 into the millions today. They run an impressive gamut: Madison, Wisconsin, boasts three hundred public bikes; Hangzhou, China, has more than sixty-six thousand.

Bicycle ridership in China followed the same trajectory as in the United States, only in a much shorter time frame and on a much larger scale. When the Communist Party seized power in 1949, it proclaimed bicycle production a national priority. The inexpensive, efficient, easily made machines were an excellent means to literally mobilize millions of workers and peasants. Beijing consolidated the country's small, private manufacturers into a few large, state firms and spoon-fed them generous quantities of rationed materials. By the late 1980s, China was producing some forty-one million bikes annually. The streets of Chinese cities soon filled with cycles, far outnumbering motor vehicles.

But automobiles would not be kept at bay. In the 1990s, China's booming economy propelled millions of its citizens up into the middle class, and the new bourgeoisie wanted cars. The government shifted support to the auto industry, seeing it as another front in the nation's march to modernity. Today, China produces more cars than any country besides Japan and the United States.

A massive shift in China's transportation patterns followed. As recently as 1986, two thirds of Beijing residents got around by bike. Today, the figure is around 15 percent. The impact of all those additional cars has been catastrophic. China's capital now has some of the most polluted air on Earth. Traffic chokes the streets of major cities across the country. Hundreds of thousands are killed in road crashes each year.

All of which helps explain why China is now, in a way, at the vanguard of a new bicycle revolution. Today, China is home to the world's most expansive bike-sharing market. (It also manufactures almost all of the sixteen million bikes imported into the US each year.) In 2017 alone, the Chinese bike-sharing industry's revenues surged tenfold from $181 million to $1.52 billion, and the number of users grew to 209 million. If there's anywhere on the planet where bikes can make a difference, it's in the densely populated, fast-growing cities of this nation of 1.4 billion people. (China is further diminishing the need to own a private car by continuing to expand what is already, by far, the world's most extensive high-speed train network. Trains traveling as fast as 200 miles per hour zip along more than 23,500 miles of track connecting all of the nation's biggest cities. All of that was built from scratch starting in 2008.)

Bicycles have their shortcomings, of course. For most people, a major one is that it takes effort to ride a bike. You don't just sit in a comfy seat and press a gas pedal to get around town. You have to expend energy, huffing and puffing your way over whatever terrain you encounter. But in the last few years, rapidly advancing technology has begun to mitigate that problem by adding the most significant change to bicycles in more than a century: electric batteries. With an electric motor to assist, or completely replace, a rider's pedaling, elderly and otherwise enfeebled people—not to mention those of us who are just lazy—can use bikes to get almost anywhere they can by car. Most trips taken in cities are only a mile or two, a distance easily traversed on an e-bike.

E-bikes are still a niche product in much of the rich world, but that's changing fast. Sales are rising as e-bikes get faster, more reliable, and cheaper. Improving battery technology and economies of scale are driving down the price of more powerful batteries. In the United States, sales of e-bikes have more than tripled since 2019, and as of 2021 were approaching one million per year. Some five million were sold in Europe that year. Many European governments are actively promoting the machines. France, for example, offers direct subsidies to people who swap a car for an e-bike.

In the developing world, writes *New York Times* climate journalist David Wallace-Wells, "the electric vehicle revolution is taking a very different shape—often with two or three wheels rather than four." There are ten times more electric scooters, mopeds, and motorcycles on the road than electric cars, he notes, "responsible already for eliminating more carbon emissions than all the world's four-wheel E.V.s." In China, by far the world's leading producer and consumer of e-bikes, they outnumber cars. More than two hundred million electric bikes are already on the roads, and thirty million new ones are sold each year. When you add in electric scooters and other micromobility devices, the worldwide market for electric vehicles smaller than cars may hit $300 billion by 2030.

But electric bikes bring us back to metals. Bicycles themselves are made of metals, and the lithium-ion batteries that power e-bikes are essentially just smaller versions of the ones that power electric cars. E-bikes also require electricity from the copper-based grid. So, even if we replaced every car on the road with an e-bike, we'd still need critical metals.

The big difference, however, is that we wouldn't need nearly as much. Bike batteries are much smaller than those built for cars and are thus built with commensurately smaller quantities of critical metals. Plus, they consume much less electricity. To take an extreme example, the lithium used in the colossal battery that powers General Motors' Hummer EV is enough to build 240 electric-bike batteries (or three smaller cars!).

Of course, bicycles by themselves can't meet all our transportation needs. They're not a substitute for reliable public transit systems. In fact, bikes work best in conjunction with public transit, providing a means to get to and from rail and bus lines. You can see that principle put into practice in places like the enormous bike-parking garage at Amsterdam's Centraal Station.

That, however, points up a problem: Half of all Americans—more than 150 million people—live in suburbs, which are typically short on public transportation. It's hard to get anywhere in these spread-out neighborhoods without a car. At least 90 million other Americans, however, live in cities—dense urban agglomerations where a car shouldn't be absolutely necessary. Worldwide, some 4.4 billion people, the majority of all human beings, live in cities. That's where we need to focus our efforts.

To make urban life without a car practical, we need to adopt several strategies. We need to make bicycling and walking easier and safer. We need to increase the amount of public transport and make it more accessible. We need, in short, to redesign our cities, to orient them around human beings rather than automobiles. In many cities all around the world, that's starting to happen.

The design of modern city streets is another aspect of life we tend to take for granted. It just seems part of the natural order of things. In the same way that it seems self-evident that forests are full of trees, it seems self-evident that cities are full of big roads dense with automobiles. But cities are not forests. They didn't grow up organically. Like everything else made by humans, they are the result of deliberate choices, of decisions made, influence applied, and interests served. For most of the last one hundred–plus years, they have been consciously shaped to serve the automobile.

"In the early 1900s there was a cross-over period when there was no dominant mode of transport in the cities of America. Pedestrians, cyclists, equestrians and motorists, all shared the usually ill-defined roads," writes transportation journalist Carlton Reid in *Forbes* magazine. Photos from that era show pushcart vendors, kids selling newspapers, pedestrians, horse-drawn carriages, and motorcars all jostling for position on city streets. But as the number of cars on those streets grew, so did the number of crashes they caused. In the four years following World War I, more Americans were killed by automobiles than had died fighting on the battlefields of Europe. "The public was particularly horrified by the number of children killed by car drivers. In 1925, fully one third of all traffic deaths were children, and half of those kids were killed on their home blocks," writes journalist Clive Thompson in a Medium post. It was clear something had to be done.

One suggested remedy for this carnage was to force cars to slow down, either by fitting their engines with so-called speed governors or equipping streets with bollards to impede their movement. That did not sound good to the automobile and oil industries. Half the reason their customers wanted cars was because they were fast. If city streets forced drivers to slow down, sales would suffer. Big Car preferred the approach put forward by Edward J. Mehren, a noted roadbuilding engineer: Don't change the cars, change the streets. In a 1922 editorial, Mehren called for "a radical revision of our conception of what a city street is for." Those streets, Mehren declared, should be reengineered to get everything else out of the way so that cars could move more quickly and easily. Mehren's ideas, promoted with lobbying and advertising campaigns from pro-car interests, won the day.

City after city widened roads, built parking areas, and restricted where pedestrians could go. According to Peter Norton in *Fighting Traffic: The Dawn of the Motor Age in the American City*, "By 1930, most streets were primarily motor thoroughfares where children did not belong and where pedestrians were condemned as 'jaywalkers,'" a criminal class that had only recently been invented.

But after a century of auto dominance, the balance of power is beginning to shift again. In the decades since Vic Langenhoff, the angry father in the Netherlands, helped kick off the movement to make city streets more hospitable to humans, it has slowly but steadily spread around the world. That movement has been turbocharged by concern about climate change and air pollution; fewer cars mean less of both.

Think about what midtown Manhattan looked like back in 1971—the snarled traffic, jostling yellow cabs, *Midnight Cowboy*'s Ratso Rizzo banging on the hood of a car and yelling "I'm walkin' here!" Today, stretches of Broadway are closed to cars. Bike lanes line the thoroughfares. Tourists lounge at cafe tables in the middle of Times Square. That's largely because the ideas behind Langenhoff's movement were put into practice in the 2010s by New York City transportation commissioner Janette Sadik-Khan. If those ideas can make it there, they can make it anywhere. Today, hundreds of cities, large and small, are redesigning their streets and rethinking their approaches to traffic, parking, and other issues to make their streets more accessible to bicycles and pedestrians and less inviting for cars.

I can tell you from personal experience that this kind of strategy works. I lived for many years with my wife and two kids in Los Angeles, where we had two cars (including that Leaf). We needed them because it's pretty much impossible to get around LA without a car. In 2020, we moved to Vancouver, Canada (leaving behind the Leaf). There, we need only one car. That's because, over the last couple of decades, Vancouver has been retrofitted into a place where you can get around easily and safely without needing a motor vehicle. The city is crisscrossed with physically protected bike lanes and streets where car traffic is limited. It has a decent bus system and an excellent light-rail system. Most of the time, the four of us get to our various destinations on foot, bikes, and public transit.

We do use our one car. We take it when the weather is bad, or when someone has to travel a long distance or carry a heavy load. (I don't recommend hauling groceries for a family of four on a bike—though even that can be done with an electric cargo model.) Bicycles will never completely replace automobiles, but if reconfigured cities can convert a significant number of car trips into bike trips, we will reap huge benefits.

Perhaps the main obstacle to using bikes, whether electric or human-powered, for transportation is that riding on most city streets around the world—swarming with two-ton metal automobiles piloted by distractable drivers who could inadvertently mow a cyclist down without scratching their fenders—is just plain dangerous. As Amsterdam shows, however, that problem can be fixed. In 1975, the per capita rate of fatal car crashes in the Netherlands was higher than that of the United States. Today, it's less than half.

By now, enough places have tried enough solutions that we have a pretty clear idea of what works. Bikes need to be separated from cars by more than just painted lines. They need a physical barrier, like a concrete curb, or even whole streets dedicated exclusively to two-wheeled transport. Those pathways need to connect to each other and form a continuous network that allows cyclists to move around the city. When those changes are made, bicycles follow.

New York City has added hundreds of miles of bike lanes to its streets since 2000 and has seen the number of bicycle commuters more than triple. In London, after several years of concerted effort to limit cars in the central city, bicycles now outnumber automobiles, and carbon emissions and pollution levels have plummeted. In Seville, Spain, the addition of seventy-five miles of protected bike lanes in the 2010s sparked an elevenfold increase in cycling. Paris has announced plans to double its amount of cycling lanes to 1,400 kilometers' worth, and create ten thousand bike-parking spots while slashing the number of car-parking spots. That's part of an official push to make the City of Light a fifteen-minute city—an urban planning concept that aims to enable residents to walk or bike to jobs, shops, transit, and other needs within a quarter-hour. Already, improved bicycling and public transportation in France's capital have cut car use in half since 1990.

Some cities are even outright banning cars in certain areas. Dozens of European cities, including London and Paris, have designated car-free zones and streets, just like those stretches of Broadway in New York City. Many have also imposed some version of what's known as congestion charging, requiring drivers to pay to drive in particularly crowded zones in the hopes that the cost will keep their numbers down.

Finland's capital, Helsinki, aims to completely eliminate the need to own a private car in the coming years. The goal is to integrate public transport, shared bikes, and ride-hailing services through a mobile phone app. Users will input their location and destination, and the app will map out a multimodal route—walk to this bus stop, take the bus to this bike-share kiosk, then ride the last three blocks.

This approach is known as mobility as a service. The notion is that, in the same way that we don't all need to own our own chainsaw or rug shampooer, we don't all need to own our own cars. Chainsaws are useful (and also quite fun). Rug shampooers are useful (though not as much fun). But most of us don't buy these things and keep them around for the rare occasions when we'll need them. Why would you invest that much cash and use up that much storage space on a machine you use only once in a while? When we need them, we rent them, borrow them, or hire someone who owns one to put them to use for us.

That's the principle behind ride-hailing services like Uber and Lyft. In practice, however, those businesses have often made traffic worse. Many of their customers are people who otherwise would have taken public transportation, not driven their own cars. Studies have shown that Uber has actually increased the number of cars and worsened traffic in many downtowns. Still, properly regulated, this type of service makes all kinds of sense.

Taken together, the upsurge in bicycling, improved public transit, and expanded ride-hailing is clearly convincing a growing number of Americans, especially younger ones, that they don't actually need their own cars. In the late twentieth century, when I was growing up, every kid ran to take their driver's test the minute they were old enough. Getting a license was a powerful rite of passage and a declaration of independence from parents. The Supreme Court even proclaimed, in 1977, that having a car was a "virtual necessity" for Americans.

Apparently, that's no longer the case. The number of young Americans getting driver's licenses is plummeting. In 1997, 43 percent of sixteen-year-olds had their licenses. By 2020, that number had dropped to barely 25 percent. The fall is most dramatic among teens, but the proportion of people with licenses has slipped in every age group under forty. In 1983, the share of Americans between twenty and twenty-four that did not have a license was one out of twelve; today, it's one out of five.

Those drops are driven by many factors. Thanks to the internet, teens don't have to drive to a mall to buy sneakers, let alone anachronisms like CDs or video-game cartridges. In many places, it's also more difficult and takes longer to get a license than it used to. But research clearly shows that the rise of alternative means of transport and concerns about climate change are encouraging young people to go car-free.

As always, there are costs. A widespread shift away from private-car ownership will cost jobs in the automotive industries and their allied businesses, like gas stations. Technological and social changes of this magnitude always entail big economic dislocations. Think of all the blacksmiths and stablehands who lost their jobs when the car replaced the horse. Ideally, displaced autoworkers will find new jobs created by the Electro-Digital Age, such as installing and maintaining renewable-energy systems or repairing and recycling electronics. "These are jobs that can't be offshored to China or Mexico or even done by robots," writes Saul Griffith in *Electrify*. "These are jobs in every zip code in America. These are skilled blue- and white-collar jobs, the great majority in the trades—electrical, plumbing, and construction—that will pay well."

In any case, if such a shift really happens on a major scale, the auto industry will most likely shrink, not disappear. We will still need cars. And some people will just *want* cars, needed or not. That's fine, especially if those cars are electric. If you want to live in a low-density exurb and get around in an SUV, that's your right. That said, it's well past time for car owners to start paying the real price for their vehicles. As it is today, every American taxpayer, regardless of how they get from place to place, helps foot the bill to keep motor vehicles moving.

The rise of automobiles was supported at every step by lavish public subsidies. The industry would never have taken off were it not for the countless billions of dollars of government money that were poured into building America's network of public roadways, including the vast interstate highway system. "Taxpayers subsidize motor vehicles with free parking spaces, roadwork, bridges, traffic enforcement, [and] environmental cleanup," writes Peter Dauvergne in *The Shadows of Consumption*. He also estimates that the total cost of traffic accidents, "from medical treatment to insurance to disability to police and legal services," exceeds $230 billion.

Add to all that the stupendous costs of subsidizing the fuels that power internal combustion engines. The International Monetary Fund estimates the United States spent $649 billion in subsidies for fossil fuels in 2017 alone. Worldwide, the figure was $5.2 trillion. Americans can also add the staggering expense of military bases in the Persian Gulf and elsewhere, not to mention the occasional Middle Eastern war, as government spending intended to keep oil flowing.

In principle, this kind of public financial support (excluding the military aspects) isn't a bad thing. Pouring public money into a particular endeavor is not only sensible but crucial when the public's safety, even survival, is at stake. We spent huge sums building electrical infrastructure in the 1920s. We spent countless billions developing a COVID-19 vaccine. Big subsidies to particular industries often make a lot of sense. Continuing to subsidize fossil fuels in a world imperiled by climate change, however, does not.

To the contrary, we should impose additional taxes or charges on motor vehicles of all types, including electric ones, and use the revenue to support public transit and bicycle infrastructure. That's a very tough sell to American voters, of course, particularly people living in rural areas who depend on their vehicles. Nonetheless, it's common practice in many industrialized countries, which subject both fuel and cars themselves to higher taxes than in the United States. It's certainly more fair. As it is, cars impose costs on all of us. Why not shift more of the burden of those costs onto those who choose to use the product?

If we're going to subsidize any industrial endeavor, it should be the transition to renewable energy and electric vehicles. The free market isn't going to get us where we need to go all by itself. Policymakers have recognized this fact and are starting to act on it—or, at least, on parts of the equation. The US government's 2022 Inflation Reduction Act includes by far the biggest, most ambitious set of credits, loans, and grants ever to support the energy transition. It earmarks tens of billions of dollars to help fund solar, wind, and other renewable technologies, electric vehicle battery development, and tax credits for people who buy EVs.

But the measure mainly promotes electrification and electric cars. As Urban Institute researcher Yonah Freemark points out, it provides no direct support for public transit, biking, or walking. That reflects the US government's preference for the car-centric status quo. Another major funding bill, the 2021 Infrastructure Investment and Jobs Act, did include $1.6 billion for transit e-buses. Then again, it also included $350 billion for federal highways. Directing that kind of money to build high-speed rail lines and thousands of miles of bike lanes instead could reduce our dependency on cars.

Let me say it again: Even with the best public transit and bicycling infrastructure, we won't get rid of motorized vehicles anytime soon, if ever. Nor should we. There will always be a need and a desire for them. There's no question, however, that with the right planning and support, humanity can get by with many fewer cars. Millions of people around the world could, and gladly would, abandon the hassle and expense of driving their own cars if they could instead get from place to place quickly, easily, and safely by bike, bus, train, or foot. Every step we take to help them do that takes us all a little further down the road to a more sustainable future.

What might it be like to live in that future, in a world built around renewable energy and more sustainable lifestyles? Ideally, at least for those of us in wealthy, developed countries, our lives would not feel all that different from those we already lead. There would be some trade-offs, of course. Some things would be a little easier, some a little harder. We might have fewer material possessions but more time to enjoy ourselves. In less wealthy parts of the world, at least for some people, life might be a whole lot better.

Here's how I imagine that world might look for Elena, a typical office worker in a typical North American city living in the near future in a best-case version of the Electro-Digital Age:

Elena looks up as her smartphone starts chirping the opening bars of an old Taylor Swift song, telling her it's time to get going. "Already?" she thinks grudgingly. But she can't be late to work. The meeting this morning could deliver a career-defining deal.

She knocks back the last of her coffee, gathers up her phone and keys, and hears her apartment door lock behind her as she steps into the hallway. It's a new building, occupying a nicely located corner of a downtown-adjacent neighborhood where a gas station used to be. Elena has lived in this city for most of her life. She's constantly amazed at how many new apartments and condos are going up as gas stations and parking lots keep getting torn down due to their car-driving customer base shrinking every year. The building has an elevator, powered by a solar array on the roof, but Elena opts to trot down the stairs to the street, a quick warmup before getting on her bike.

On the ground floor, she ducks into the little clothes shop tucked into the lobby to pick up her jacket. The place sells new clothes, but, like most newer boutiques, also does repair—in her case, swapping out her busted old zipper for a sturdier new one.

Elena keys in the code to unlock the door to the bike-parking structure next to the shop. She'd still rather be upstairs on her couch than heading off to work, but at least the commute is quicker and easier than it was when she started with the company in the mid-2020s. Back then, when cars still ruled all the city's main roads, Elena used to battle her way through the traffic in her SUV. She'd chuckle over how absurd it was that, like most other commuters, she was usually the sole occupant of a vehicle designed to carry a whole family. The decision to switch had been gradual. She had started telecommuting a couple of days a week, which meant she didn't need the car as often. Meanwhile, the cost of owning it kept going up, as the government added on taxes and fees to offset the health and environmental costs motor vehicles incurred. Car manufacturers were hiking retail prices, too, to cover the expense of their newly mandated recycling responsibilities. Then the city started charging for street parking almost everywhere and for driving downtown during business hours.

Between all of that and her own ambient concern about climate change, Elena had eventually decided that it just wasn't worth owning a motor vehicle anymore. Especially since, by then, there were plenty of alternative ways to get around. Bicycles, ride-hailing services, subways, trolleys, and her own feet picked up the slack handily. The city's new subway system was still being built, and the construction sites pocking downtown were a nightmare to navigate around. Elena, though, was lucky enough to live near one of the completed lines—not quite close enough to walk, but an easy bike ride.

Elena used to consider bicycles toys for suburban tweens, spandex-clad fitness freaks, and virtue-signaling vegans, but a hefty rebate program—thanks, state legislators!—had lured her into buying an electric bike. Riding it around town, she discovered, was much easier and safer than she'd imagined. At first, she'd mostly taken it out on weekends, riding for fun with friends on a stretch of freeway that had been converted into a park, on the model of New York City's High Line. She'd quickly realized that the bike could actually replace her SUV as her main means of transportation. The thinning of the automobile herds had opened up space on the streets for smaller vehicles. Where there used to be a wall of parked cars lining the curb in front of her apartment building, there was now a bike lane shielded from cars by a low concrete barrier. On sunny days, she liked to ride all the way to the office, taking protected bike paths and low-traffic streets the whole way.

This fall morning, though, is chilly and overcast. Elena considers resorting to an Uber, as she usually does when the weather is ugly. The extra expense irritates her, but, as she always reminds herself, when she adds up all the money she saves by not owning a car—no insurance, no maintenance, no gas, no parking fees—she figures she still comes out ahead. And the rides come quickly. There are still lots of cars on the roads. Her brother Jacob, who lives out in the suburbs, had laughed in her face when she had suggested he give up his gas-guzzling minivan. "How am I supposed to get the kids everywhere? On a skateboard?" he'd snorted. "This van is the school bus, soccer-practice shuttle, and grocery-delivery vehicle. Let's see you do all that on a bicycle."

At least he had switched to an electric model. Tax credits and other incentives had helped convince him and millions of other drivers to go electric. Of course, Jacob also complained that the country couldn't afford all those tax credits. They were being phased out, however, now that economies of scale and improving technologies had made the prices of EVs competitive with their gas-fueled counterparts.

In any case, Elena's weather app tells her that today will be cold but with no rain expected. So, with the e-bike's battery doing most of the work, she rolls off to the subway station. She scrolls through the news as she waits for the train, elbows angled out to maintain a little personal space on the crowded platform. She glances at the first headline: NEW UN REPORT FINDS GLOBAL CARBON EMISSIONS CONTINUE TO DECLINE. Good news, but boring. CALIFORNIA TO BUILD HIGH-SPEED TRAIN FROM LA TO BOOMING 'LITHIUM VALLEY.' Ditto. TENSIONS HIGH AFTER NEW THREAT FROM CHINA TO BLOCKADE TAIWAN. Elena clicks that one; this is news that might affect her own industry. She nods as she reads that the simmering Taiwan crisis is prompting the US, Canada, and Europe to redouble their efforts to build critical metal supply chains outside of China's control.

More mines are being opened up worldwide, but not nearly as many as predicted back in the 2020s. The drop in car ownership has reduced the anticipated demand for critical metals, and the fast-growing recycling industry is supplying more of those materials every day.

Elena herself works in the e-waste recycling sector, somewhat to her own surprise. She'd started her corporate career years ago in the automobile-parts world, working at several different businesses along the way. But when the recycling company opened a branch in her city and started hiring aggressively, Elena made the switch. At that time, the company was growing like crazy, along with the whole industry. Several factors were driving the boom. Tax breaks and corporate payments under extended producer-responsibility laws helped. And as mines around the world were obliged to raise pay and improve conditions for their workers, the cost of virgin metals started rising. Meanwhile, the costs of recycled metals were falling, as AI-powered systems made disassembling and sorting increasingly faster and more efficient. The ongoing energy transition meant there was a huge market for recycled resources. Elena knew lots of people were shifting from the stagnating automobile and oil and gas industries into the repair, recycling, and renewable-energy sectors, and figured she should give it a try. Her skills translated pretty easily. Most of her experience was in dealing with suppliers, trying to get the best deals on raw materials for her company's factories. Recyclers needed people to do that just as much as spark-plug manufacturers did.

The meeting today, though, isn't business as usual. It's a rare opportunity for Elena and her team to try to convince one of the world's top e-waste suppliers—a multibillion-dollar corporation employing tens of thousands of people in its home country—to strike a partnership with her company.

At her office desk, Elena pulls up a slide deck on her laptop. The machine looks brand-new, though she bought it used at the local Apple retail store. Like the outlets of every major electronics retailer, Apple Stores now have whole sections selling refurbished devices at a discount. Her machine also contains certified recycled metals—some of which, Elena is always amused to think, may well have come from her own company's smelters. She'll have to remember to mention that when she pitches the suppliers in just a few minutes.

The office lights dim for a second, then regain strength—a sign that the building has just switched its power source from the city grid to the semitrailer-sized battery in the subbasement. That happens when the prairie winds in the neighboring county die down, and the battalions of wind turbines standing in a decommissioned natural-gas field stop turning. Elena barely notices. It happens all the time. She's focused on putting a final polish on her PowerPoint slides, mentally rehearsing her pitch, and psyching herself up to convince the men and women now filing into the conference room that they should cut her company in on some of the billions of dollars' worth of printed circuit boards, batteries, and other e-waste they export every year. Elena takes a deep breath, steps into the room, and introduces herself to the team from Abubakar Industries.

"Welcome!" she says. "How was the flight from Nigeria?"

The world, even in this idealized vision of the Electro-Digital Age, is still far from perfect. Renewable energy and reduced consumption aren't a cure for poverty, violence, or exploitation. They aren't a cure for climate change, either, but they are necessary components of the cure.

This is a crucial moment. We are still in the early stages of the digital revolution and at the very beginning of the shift from fossil fuels to renewables. The choices we make now about how to manage the transition into the Electro-Digital Age will have huge impacts in the decades ahead as the process grows in speed and scale. Those choices go far beyond whether or not we should be mining lithium from the Atacama Desert or cobalt from the ocean floor. We need to be looking for every opportunity to minimize the transition's costs and to maximize its benefits. Where we mine, we need to do as much as possible to safeguard the environment and the people who work in and live near those mines. Where we are dependent on China, or any other unreliable source, we need to break that dependency. We need to devote major efforts and investment into building up our capacity to recycle, repair, and reuse. All of that is doable. Some of it is already being done.

All of those measures will help ease the damaging side effects of this new era. But we still need to do more, by which I really mean *less*. We need to swap fossil fuels for renewable-energy sources, *and* we need to reduce our use of energy across the board. We need to switch from gasoline cars to electric vehicles, *and* we need to reduce the total number of cars on the world's roads. The less we consume, the less energy we need. The less energy we use, the less metal we need to dig up, the less pollution we generate, and the less dependent on foreign suppliers we are. If we are more careful about how we extract resources and more thoughtful about what and how much we consume, we have a shot at getting to the world we want without sacrificing deserts, rainforests, oceans, and human lives in the process.

Our future depends, in a literal sense, on metal. We need a lot of it to stave off climate change, the most dangerous threat of all. But the less of it we use, the better off we'll all be.

ACKNOWLEDGMENTS

Looking back at the years of research and writing that went into this book, I'm astonished by how many people helped me out along the way, in so many different ways. I'm tremendously grateful to all of them.

Early on, W. Scott Dunbar gave me an excellent overview of how mining works, including a copy of his book, *How Mining Works*. Adam Minter's *Junkyard Planet* and our conversation about the international trade in scrap metal and e-waste completely reset my thinking on those subjects and helped me frame parts of the second half of this book. My cousin John Lando also shared hours of his knowledge about mining and metals markets, and I just hope that all the beer we drank in the process made it worth his while.

When I hit the road, Jon Bartlett hooked me up like a true pro with contacts in Chile from Santiago to the Atacama Desert. Zahel Quezada got us out to copper mines and remote corners of Chile's north while keeping up a captivating political commentary the whole way. Special thanks to Muriel Alarcon, who very generously volunteered some last-minute interview translating. A hearty *hvala* to my pals Annette Mangard and Gary Popovic for hosting me in Toronto. In Nigeria, the indomitable Bukola Adebayo cheerfully guided me through mud pits, scrapyards, garbage dumps, and the most chaotic traffic on Earth to get us to parts of Lagos I could never have reached otherwise.

Many other folks generously shared their time and knowledge with me but did not end up quoted or named in the text. In that category, I am especially indebted to Kent MacWilliam, Ian Morse, Dan Crocker, Patricia Mohr, Mark Selby, Shawn Hood, Christopher Grove, Adegun Oluwatobi and Rachel "Toyosi" Ademola, Sonia Dias, Josh Lepawsky, Thea Riofrancos, Hans Eric Melin, Stephen D'Esposito, and Mel Davis.

Much gratitude to the Canada Council for the Arts, the Pulitzer Center on Crisis Reporting, the Dave Greber Freelance Writers Awards, and the Access Copyright Foundation, all of whom provided financial assistance to this project. Travel is expensive, and writing a book takes a long, mostly unpaid time. As the news business continues to crumble, these kinds of supporters are increasingly important to independent journalists like myself. Speaking of supporters of independent journalists: a heartfelt salute to Maria Streshinsky, Sandra Upson and Tom Simonite at *Wired*, Jason Mark and Paul Rauber at *Sierra* magazine, and Jennifer Turner at the Wilson Center for editing and publishing several articles that became part of this book. Friendly fellow authors Parag Khanna, J. B. MacKinnon, Charles Montgomery, and especially Taras Grescoe also chimed in with advice and moral support.

Special thanks to Lisa Bankoff, my ace literary agent, as well as Jake Morrissey, my ever-patient editor, plus Cliff Corcoran, Ashley Garland, Catalina Trigo, and the rest of the team at Riverhead Books. Thanks also to Kelsey Lannin for unfailingly good-natured research and fact-checking assistance.

I am blessed with a multifarious assortment of cousins and other relatives, many of whom helped out in their own unique ways. Jeffery Lando (and Bear and Roxie) helped me think through ideas on our frequent walks together. Jennie Kamin and Neil Gustafson provided marketing guidance. Marianne Kaplan consulted on matters South African. Sarah Lando kept me focused in our coworking sessions at the library. Dominique Lando, Brian Malis, and Juniper Maylis several times let me be a literal writer-in-residence at their house on Bowen Island. Hope to see you all in the summer!

Finally, my deepest thanks to my home team: Isaiah, Adara, and Kaile. They tolerated months—years!—of me kvetching and jabbering about battery chemistries and parking policies and other tedious topics, and still gave outstanding reader feedback, advice, support, and just the right amount of sarcasm. All my love, always.

NOTES

Dear Reader: This book covers a lot of rapidly changing terrain. I have tried to use the most up-to-date facts and figures I could find, but it's inevitable that some of the details of specific policies, technologies, markets, and plain old human behavior will have changed by the time you are reading this. It's also possible that I have made mistakes, though I certainly did my best to be as accurate as possible on all fronts. If you have a question, update, or possible correction to offer about any particular fact, please feel free to drop me a line via my website at vincebeiser.com.

Unless otherwise noted, all quotes are from interviews I conducted. I have footnoted only those facts I think might be particularly surprising, contentious, or difficult for the reader to verify on their own.

CHAPTER I: THE ELECTRO-DIGITAL AGE

2 **two thirds of all the elements:** Paul McGuiness and Romana Ogrin, eds., *Securing Technology-Critical Metals for Britain* (Birmingham, UK: University of Birmingham, 2021), 36, birmingham.ac.uk/documents/college-eps/energy/policy/policy-comission-securing-technology-critical-metals-for-britain.pdf.

2 **dozens of different metals:** Guillaume Pitron, *The Rare Metals War*, trans. Bianca Jacobsohn (London: Scribe, 2023), 44.

2 **Europium enhances the colors:** McGuiness, *Securing Technology-Critical Metals for Britain*, 96.

3 **Just one Tesla Model S:** Jens Glüsing et al., "Mining the Planet to Death: The Dirty Truth about Clean Technologies," *Spiegel International*, November 4, 2021, spiegel.de/international/world/mining-the-planet-to-death-the-dirty-truth-about-clean-technologies-a-696d7adf-35db-4844-80be-dbd1ab698fa3.

3 **coal-fired power plants:** International Energy Agency, *Renewables 2022* (Paris: IEA, 2022), 10, iea.org/reports/renewables-2022.

3 **a little more than 4 percent:** The Red Cloud farm produces 350 megawatts out of Los Angeles' total of 8,019 megawatts. Los Angeles Department of Power and

Water, "Mayor Garcetti Announces That Over 60% of LA's Energy Is Now Carbon-Free," press release, February 23, 2022; Los Angeles Department of Power and Water, "Facts and Figures, 2020–21," ladwp-jtti.s3.us-west-2.amazonaws.com/wp-content /uploads/sites/3/2021/10/04152431/2020-2021_Facts_and_Figures_Digital_final .pdf.

3 **dozens of towering turbines:** Metals in turbines etc.: The Mining Association of Canada, "Mining and Its Role in Clean Technology," mining.ca/our-focus/climate -change/mining-and-its-role-in-clean-technology. On niobium in wind turbine towers, see CBMM North America, "Steelmakers Meet Demand for Taller Wind Towers with Low Carbon Structural Steel Containing Niobium," assets.niobium .tech/-/media/NiobiumTech/Documentos/Resource-Center/NT_Taller-wind -towers-with-low-coast-steel-containing-niobium.pdf.

4 **That battery is made with:** Vaclav Smil, *How the World Really Works* (New York: Viking, 2022), 101.

4 **seven hundred million tons of copper:** US Geological Survey, "How Much Copper Has Been Found in the World?," usgs.gov/faqs/how-much-copper-has-been -found-world; S&P Global, *The Future of Copper* (New York: S&P Global, July 2022), 9, cdn.ihsmarkit.com/www/pdf/0722/The-Future-of-Copper_Full-Report _14July2022.pdf.

4 **the International Energy Agency estimates:** Many organizations and researchers have published different estimates of future critical-metal demand, and those estimates change constantly. Policy shifts, changes in market conditions, new technologies, and other factors all come into play. I've chosen the IEA's estimates because the agency is generally accepted as an authoritative source on the subject. These particular numbers were accessed on August 22, 2023, based on the Announced Pledges Scenario on the IEA's Critical Minerals Data Explorer, iea.org /data-and-statistics/data-tools/critical-minerals-data-explorer.

4 **lithium, fifteen times higher:** Note this is lithium, not lithium carbonate equivalent, which is about five times as much.

4 **The market size:** International Energy Agency, *Critical Minerals Market Review 2023* (Paris: IEA, 2023), 12, iea.org/reports/critical-minerals-market-review-2023.

4 **"Energy transition minerals":** International Energy Agency, *Critical Minerals Market Review 2023*, 5.

4 **The market for critical metals:** International Energy Agency, *World Energy Outlook 2022* (Paris: IEA, 2022), 217, iea.org/reports/world-energy-outlook-2022.

5 **"The prospect of a rapid increase":** International Energy Agency, *The Role of Critical Minerals in Clean Energy Transitions* (Paris: IEA, 2021), 13, iea.org/reports /the-role-of-critical-minerals-in-clean-energy-transitions.

5 **America's leading toxic polluter:** US Environmental Protection Agency, "Metal Mining," March 2023, epa.gov/trinationalanalysis/metal-mining.

5 **sullied the watersheds of almost half:** Jared Diamond, *Collapse* (New York: Viking, 2005), 14.

5 **Torrents of toxic sludge:** The worst such dam failure in US history was in West Virginia in 1972. It killed 125 people. That was eclipsed in 2019 by a mining-dam collapse in Brazil that polluted nearly two hundred miles of rivers and killed 270 people.

5 **Those casualties are on top of:** Responsible Mining Foundation, *Harmful Impacts of Mining* (Canton de Vaud: RMF, 2021), 25, responsibleminingfoundation.org /app/uploads/RMF_Harmful_Impacts_Report_EN.pdf.

6 **A study from the Vienna University:** Stefan Giljum et al., "A Pantropical Assessment of Deforestation Caused by Industrial Mining," *PNAS* 119, no. 38 (September 12, 2022), pnas.org/doi/full/10.1073/pnas.2118273119.

6 **Seventy-five pounds of ore:** Brian Merchant, "Everything That's Inside Your iPhone," *Vice*, August 15, 2017, vice.com/en/article/433wyq/everything-thats-inside -your-iphone.

6 **a single four-and-a-half-ounce iPhone:** Pitron, *The Rare Metals War*, 44.

6 **as much as one hundred pounds:** Aaron Perzanowski, *The Right to Repair* (Cambridge: Cambridge University Press, 2022), 34.

6 **huge quantities of water:** Perzanowski, *The Right to Repair*, 31.

6 **as much as 7 percent:** Lindsay Delevingne et al., "Climate Risk and Decarbonization: What Every Mining CEO Needs to Know," *McKinsey Sustainability*, January 28, 2020, mckinsey.com/capabilities/sustainability/our-insights/climate-risk-and -decarbonization-what-every-mining-ceo-needs-to-know.

6 **At least 320 antimining activists:** Global Witness, "1910 Land and Environmental Defenders Were Killed between 2012 and 2022," globalwitness.org/en /campaigns/environmental-activists/numbers-lethal-attacks-against-defenders -2012.

7 **an estimated 5.35 billion people:** Ani Petrosyan, "Number of Internet and Social Media Users Worldwide as of January 2024," Statista, January 31, 2024, statista.com/statistics/617136/digital-population-worldwide.

7 **Nearly as many own mobile phones:** Bruno Venditti, "This Graphic Shows What Your Smartphone Is Made of," World Economic Forum, August 27, 2021, weforum.org/agenda/2021/08/this-visualization-breaks-down-the-metals-in-a -smartphone.

7 **fifteen billion mobile devices:** Frederica Laricchia, "Forecast Number of Mobile Devices Worldwide from 2020 to 2025," Statista, March 10, 2023, statista.com /statistics/245501/multiple-mobile-device-ownership-worldwide.

8 **imperiling the hydro-energy supply:** Sharon Bernstein, Jake Spring, and David Stanway, "Insight: Droughts Shrink Hydropower, Pose Risk to Global Push to Clean Energy," Reuters, August 13, 2021.

8 **more than 12 percent:** Ember, *Global Electricity Insights*, ember-climate.org /insights/research/global-electricity-review-2024/global-electricity-source-trends/; BP, *BP Statistical Review of World Energy 2022* (London: BP, 2021) 3; Nathaniel Bullard, "Four Charts Reveal Seismic Shifts in Global Energy Within One Lifetime," *Bloomberg*, June 30, 2022.

8 **a quarter of the nation's electricity:** US Energy Information Administration, "Short-Term Energy Outlook," accessed January 4, 2024, eia.gov/outlooks/steo.

8 **The IEA expects:** International Energy Agency, *Renewables 2022*, 30.

8 **"unprecedented momentum for renewables":** International Energy Agency, *Renewables 2022*, 10.

8 **In 2023, the world created:** International Energy Agency, *Renewables 2023* (Paris: IEA, 2024), 8–9, iea.org/reports/renewables-2023.

8 **By 2027, the IEA predicts:** International Energy Agency, *Renewables 2022*, 10, 26.

8 **only 120,000 electric vehicles:** International Energy Agency, *Global EV Outlook 2022* (Paris, IEA, 2022), 5, iea.org/reports/global-ev-outlook-2022.

8 **By 2022, customers were snapping up:** David Wallace-Wells, "Electric Vehicles Keep Defying Almost Everyone's Predictions," *The New York Times*, January 11, 2023.

8 **Sales of new electric vehicles:** NREL, "Building the 2030 National Charging Network," nrel.gov/news/program/2023/building-the-2030-national-charging-network.html; Ian Telfer and Patricia Mohr, "Commodities and Financial Markets" (presentation, AME Roundup 2022, Vancouver, BC, February 1, 2022).

8 **at least twenty countries have announced:** Henry Sanderson, *Volt Rush: The Winners and Losers in the Race to Go Green* (London: Oneworld, 2022), 11.

9 **recently sank $650 million:** Danny Lee, David Stringer, and Jacob Lorinc, "Shortage of Metals for EVs Is Rising up the Agenda in Automakers' C-Suites," *Bloomberg*, March 3, 2023, bloomberg.com/news/articles/2023-03-04/shortage-of-metals-for-ev-batteries-now-a-key-concern-for-automakers-c-suites.

9 **investing heavily in critical-metal production:** Benchmark Mineral Intelligence, "From Oil to Lithium: How Saudi Arabia Is Building a Battery Supply Chain," June 16, 2023, source.benchmarkminerals.com/article/from-oil-to-lithium-how-saudi-arabia-is-building-a-battery-supply-chain.

9 **Brazil produces almost all:** US Geological Survey, *Mineral Commodity Summaries 2022* (Reston, VA: US Geological Survey, January 2022), 116–17.

9 **world's top producer of high-grade nickel:** International Energy Agency, *World Energy Investment 2022* (Paris: IEA, 2022), 116, iea.org/reports/world-energy-investment-2022.

9 **sent the metal's prices skyrocketing:** International Energy Agency, *World Energy Investment 2022*, 117, 128.

10 **Though most Russian exports:** Eric Onstad, "EU, US Step Up Russian Aluminum, Nickel Imports since Ukraine War," Reuters, September 7, 2022.

10 **Leveraging its own natural resources:** The White House, *Building Resilient Supply Chains, Revitalizing American Manufacturing, and Fostering Broad-Based Growth*, 100-Day Reviews under Executive Order 14017 (Washington, DC: The White House, June 2021), 121.

10 **China has huge reserves of lithium:** International Energy Agency, *World Energy Investment 2022*, 130–31.

10 **allegedly mines with forced labor:** Anna Swanson and Chris Buckley, "Red Flags for Forced Labor Found in China's Car Battery Supply Chain," *The New York Times*, November 4, 2022.

10 **most will end up sent to China:** International Energy Agency, *World Energy Investment 2022*, 130–31.

10 **China has more than half:** International Energy Agency, *Global EV Outlook 2022*, 7–8; Benchmark Mineral Intelligence, "Infographic: China's Lithium Ion Battery Supply Chain Dominance," October 3, 2022; Steve LeVine, "America Isn't Ready for the Electric-Vehicle Revolution," *The New York Times*, November 10, 2021.

10 **close to that much for nickel and copper:** Dolf Gielen, *Critical Materials for the Energy Transition* (Abu Dhabi: International Renewable Energy Agency, May 2021), 16; S&P Global, *The Future of Copper*, 12.

10 **most of the world's solar panels:** Daniel Yergin, *The New Map* (New York: Penguin Press, 2020), 341, 396–97.

10 **hefty share of its wind turbines:** Shashi Barla, *Global Wind Turbine OEMs 2020 Market Share* (Edinburgh, UK: Wood Mackenzie, March 31, 2021), 3.

10 **nearly three quarters of all lithium-ion batteries:** Yergin, *The New Map*, 341.

10 **majority of all electric vehicles:** Daisuke Wakabayashi and Claire Fu, "For China's Auto Market, Electric Isn't the Future. It's the Present," *The New York Times*, September 27, 2022.

11 **pouring cash and resources into the quest:** International Energy Agency, *Critical Minerals Market Review 2023*, 6.

11 **"The United States' mineral import dependency":** S&P Global, *The Future of Copper*, 15.

11 **Congress enacted an infrastructure package:** The White House, "Fact Sheet: Biden–Harris Administration Driving US Battery Manufacturing and Good-Paying Jobs," October 19, 2022, whitehouse.gov/briefing-room/statements-releases /2022/10/19/fact-sheet-biden-harris-administration-driving-u-s-battery -manufacturing-and-good-paying-jobs.

11 **passed the Inflation Reduction Act:** US Department of Energy Alternative Fuels Data Center, "Electric Vehicle (EV) and Fuel Cell Electric Vehicle (FCEV) Tax Credit," August 16, 2022, afdc.energy.gov/laws/409.

11 **many more billions of dollars:** S&P Global, *Inflation Reduction Act: Impact on North America Metals and Minerals Market* (New York: S&P Global, August 2023), 6, 10–11, 86.

12 **"Every one of these hand phones":** From Friedland's Day 3 Keynote Presentation at the CRU World Copper Conference 2022 in Santiago, Chile, March 30, 2022.

12 **Indigenous people didn't even have:** "Mapping the Legal Consciousness of First Nations Voters: Understanding Voting Rights Mobilization; A Brief History of First Nations Voting Rights," Elections Canada, August 27, 2018, elections.ca/content .aspx?section=res&dir=rec/part/APRC/vot_rights&document=p4&lang=e.

13 **restitutions to an Aboriginal group:** "Rio Tinto Reaches Historic Agreement with Juukan Gorge Group," Reuters, November 27, 2022, reuters.com/world/asia -pacific/rio-tinto-reaches-historic-agreement-with-juukan-gorge-group -2022-11-28.

13 **"We need to let people know":** Libby Sharman (speech at AME Roundup 2023, Vancouver, Canada, January 23, 2023).

14 **"Cleaning up pollution":** Diamond, *Collapse*, 8–11.

14 **All those regulations and public hearings:** "The Transition to Clean Energy Will Mint New Commodity Superpowers," *The Economist*, March 26, 2022; S&P Global, *The Future of Copper*, 13.

14 **In the 1950s:** S&P Global, *The Future of Copper*, 69–70, 73.

15 **the International Energy Agency warned:** International Energy Agency, *World Energy Investment 2022*, 134–35.

15 **That's the 2016 treaty:** United Nations Framework Convention on Climate Change, "The Paris Agreement," unfccc.int/process-and-meetings/the-paris-agreement.

15 **"could derail or delay the energy transition":** S&P Global, *The Future of Copper*, 15.

CHAPTER 2: THE ELEMENTAL SUPERPOWER

21 **one September morning in 2010:** I have reconstructed this incident from details from several different sources. As of July 2023, video footage of much of the action

could be found at Stingmews, "China Secret Ship Attacks Japan Coast Guard 1," Dailymotion.com, dailymotion.com/video/x2s5bs6; and Waiqueure, "Sengoku 38," YouTube.com, Nov. 5, 2010, youtube.com/watch?v=lDLKkiqiVs8.

Other important sources include: Alexis Pedrick and Lisa Berry Drago, "Rare Earths: Hidden Cost to Their Magic," *Distillations* podcast, June 25, 2019, produced by Mariel Carr, Rigoberto Hernandez, and Alexis Pedrick for Science History Institute, sciencehistory.org/stories/distillations-pod/rare-earths-the-hidden-cost -to-their-magic; David Abraham, *The Elements of Power* (New Haven, CT: Yale University Press), 22–23; Wei Tian, "Arrest Brings Calamity to Trawler Captain's Family," *China Daily*, September 13, 2010; Yomiuri Shimbun, "Video Shows Clear Hits on JCG Boats," *Asia One News*, November 6, 2010.

21 **Japan took over the islands:** Council on Foreign Relations, "China's Maritime Disputes, 1895–2023," cfr.org/timeline/chinas-maritime-disputes.

21 **thirty-eight Chinese fishing ships:** Andrew H. Malcolm, "Japanese-Chinese Dispute on Isles Threatens to Delay Peace Treaty," *The New York Times*, April 15, 1978.

22 **two nations struck an agreement:** Alexis Pedrick and Elisabeth Berry Drago, "Rare Earths: The Hidden Cost to Their Magic: Parts 1 and 2," June 25, 2019, *Distillations* podcast, produced by Rigoberto Hernandez, podcast, 57:04, sciencehistory .org/distillations/podcast/rare-earths-the-hidden-cost-to-their-magic#transcript.

22 **Chinese premier Wen Jiabao:** Keith Bradsher, "Amid Tension, China Blocks Vital Exports to Japan," *The New York Times*, September 22, 2010.

22 **Chinese mines supplied 95 percent:** Keith Bradsher, "Rare Earths Stand Is Asked of G-20," *The New York Times*, November 5, 2010.

23 **"all thirty-two of the country's exporters":** Abraham, *The Elements of Power*, 24.

23 **Rare earth prices shot up:** Abraham, *The Elements of Power*, 25.

23 **fifty-nine renewable energy companies:** Julie Michelle Klinger, *Rare Earth Frontiers* (Ithaca, NY: Cornell University Press, 2018), 155.

23 **unnerved business executives:** Bradsher, "Rare Earths Stand Is Asked of G-20."

23 **The House Armed Services Committee:** Bradsher, "Amid Tension, China Blocks Vital Exports to Japan."

23 **"The world's reliance on Chinese rare earth materials":** Sophia Kalantzakos, *China and the Geopolitics of Rare Earths* (Oxford, UK: Oxford University Press, 2018), 23.

23 **Zhan was sent home:** Austin Ramzy, "Japan Releases Chinese Captain, but Tensions Remain," *Time*, September 27, 2010.

23 **all began concerted efforts to find and develop:** Kalantzakos, *China and the Geopolitics of Rare Earths*, 14, 16–17, 21–22.

24 **"They enable both the hardware and the software":** Klinger, *Rare Earth Frontiers*, 16.

24 **a German chemist figured out:** Klinger, *Rare Earth Frontiers*, 61.

25 **Billions of these mantles:** Klinger, *Rare Earth Frontiers*, 46.

25 **countries where the rare earths were mined:** Klinger, *Rare Earth Frontiers*, 62.

25 **"At the time, these costs were hidden":** Pedrick and Drago, "Rare Earths: Hidden Cost to Their Magic."

25 **prospectors in the high-desert scrublands:** Kalantzakos, *China and the Geopolitics of Rare Earths*, 2.

25 **ruptured dozens of times:** Klinger, *Rare Earth Frontiers*, 71. See also: MP Mate-

rials Corp., Annual Report (Form 10-K), US Securities and Exchange Commission, December 31, 2020, 30–31; Perzanowski, *The Right to Repair*, 31; Marla Cone, "Desert Lands Contaminated by Toxic Spills," *Los Angeles Times*, April 24, 1997.

26 **Smartphones are jammed:** Sareena Dayaram, "Metals Inside Your iPhone Are More Precious Than You Thought: Here's Why," CNET.com, November 23, 2022; Pitron, *The Rare Metals War*, 195.

26 **Many military technologies:** The Interagency Task Force in Fulfillment of Executive Order 13806, *Assessing and Strengthening the Manufacturing and Defense Industrial Base and Supply Chain Resiliency of the United States* (Washington, DC: Department of Defense, September 2018), 32.

26 **the number-one product:** Ryan Castilloux, "Rare Earths: Small Market, Big Necessity," Adamas Intelligence, 2019, 10.

26 **Tiny versions of these magnets:** Pitron, *The Rare Metals War*, 21, 194.

26 **A single wind turbine:** Keith Veronese, *Rare: The High-Stakes Race to Satisfy Our Need for the Scarcest Metals on Earth* (Buffalo, NY: Prometheus, 2015), 42.

26 **"The day after the deal":** Kalantzakos, *China and the Geopolitics of Rare Earths*, 4.

27 **developing countries—especially China:** "The World Bank in China," The World Bank, April 20, 2023, worldbank.org/en/country/china/overview.

27 **"Things began to change in the 1970s":** Sandra Atchison, "The Death of Mining [in America]," *BusinessWeek*, December 17, 1984.

27 **created the Environmental Protection Agency:** "The Origins of EPA," US Environmental Protection Agency, June 5, 2023, epa.gov/history/origins-epa.

27 **Nixon's brother Edward:** Klinger, *Rare Earth Frontiers*, 127.

27 **"In the last two decades of the twentieth century":** Pitron, *The Rare Metals War*, 68–69.

28 **one third of all the world's rare earths:** US Geological Survey, *Mineral Commodity Summaries 2023* (Reston, VA: US Geological Survey, January 2023), 142–43.

28 **"Improve the development and application":** Kalantzakos, *China and the Geopolitics of Rare Earths*, 1.

28 **"The Middle East has oil":** Ernest Scheyder, "American Quandary: How to Secure Weapons-Grade Minerals without China," Reuters, April 22, 2020.

28 **the area is the biggest hub:** Kalantzakos, *China and the Geopolitics of Rare Earths*, 3.

29 **"Every ton of rare earth":** Klinger, *Rare Earth Frontiers*, 70, 75, 134.

29 **polluted the nearby Yellow River:** Pitron, *The Rare Metals War*, 31; Klinger, *Rare Earth Frontiers*, 136.

29 **"Twenty minutes outside of Baotou":** Perzanowski, *The Right to Repair*, 32.

29 **China mines most of the world's rare earth:** International Energy Agency, *World Energy Investment 2022*, 130–31.

29 **"China represents a significant":** Interagency Task Force, *Assessing and Strengthening*, 33, 40–41.

30 **Beijing has threatened:** Jeff Pao, "China Takes Rare Earth Aim at Raytheon and Lockheed," *Asia Times*, February 22, 2022.

30 **China restricted exports:** Amy Lv and Brenda Goh, "Beijing Jabs in US-China Tech Fight with Chip Material Export Curbs," Reuters, July 4, 2023; Nick Carey, "China Gallium Curbs Raise Chip Questions for Future EV Models," Reuters, July 11, 2023. In October 2023, China also imposed new restrictions on graphite

exports. See, for instance: "China curbs graphite exports in latest critical minerals squeeze," Reuters, October 20, 2023.

30 **"In this critical minerals and materials context":** Joe Deaux, "China's Grip on Critical Minerals Draws Warnings at IEA Gathering," *Bloomberg*, September 28, 2023.

CHAPTER 3: THE GLOBAL TREASURE HUNT

32 **acquired the shuttered site:** MP Materials Corp., 2020 Annual Report, 11.

33 **that number topped forty-two thousand tons:** MP Materials Corp., Annual Report, 95; MP Materials, "MP Materials Reports Fourth Quarter and Full Year 2022 Results," press release, February 23, 2023.

33 **nearly 15 percent of all production worldwide:** US Geological Survey, *Mineral Commodity Summaries 2023*, 142–43.

33 **"$527 million in revenue":** MP Materials, "Fourth Quarter and Full Year 2022 Results."

33 **"We believe we can generate":** MP Materials Corp., Annual Report, 7.

33 **The Department of Defense has pledged:** MP Materials Corp., Annual Report, 21.

34 **one million tons of explosives:** US Geological Survey, "Advance Data Release of the 2020 Annual Tables," November 1, 2022.

35 **Earth-moving machines then load:** You can see the whole process in an MP Materials–produced video at: MP Materials, "Rare Earth Magnets—HOW They're Made," YouTube.com, March 15, 2023, youtube.com/watch?v=vXqOcZDNSfg.

36 **The worst such dam failure:** Diamond, *Collapse*, 16; Mine Safety and Health Administration, "Buffalo Creek Mine Disaster 50th Anniversary," US Department of Labor, msha.gov/buffalo-creek-mine-disaster-50th-anniversary.

36 **That was eclipsed in 2019:** Responsible Mining Foundation, *Harmful Impacts of Mining*, 17.

37 **is also building a plant in Texas:** Mary Hui, "Lynas Is Shaking Up the Supply Chain for Rare-Earth Metals," *Quartz*, March 6, 2021.

39 **recent years in Canada:** Shane Lasley, "A North of 60 Rare Earths Supply Chain," *North of 60 Mining News*, December 2, 2021.

39 **especially in Australia:** Pitron, *The Rare Metals War*, 61.

39 **angry locals and legislators thwarted:** "Greenland Minerals' Kvanefjeld Rare-Earths Project Hits Roadblock," *Mining Technology*, September 27, 2022.

39 **Protesters, including Greta Thunberg:** "Sweden Gives Green Light to Controversial Iron Mine," *Deutsche Welle*, March 22, 2022.

39 **may be forced to close:** Melanie Burton, "Lynas' Malaysia Rare Earths Plant Faces Part Closure as Regulator Keeps Curbs," Reuters, February 13, 2023.

40 **to the neighboring nation of Myanmar:** Dake Kang, Victoria Milko, and Lori Hinnant, "'The Sacrifice Zone': Myanmar Bears Cost of Green Energy," Associated Press, August 9, 2022.

40 **In 2014, Myanmar exported:** "Myanmar's Poisoned Mountains," Global Witness, August 9, 2022, globalwitness.org/en/campaigns/natural-resource-governance/myanmars-poisoned-mountains.

40 **that figure had rocketed to $780 million:** "China's Rare Earth Imports from Myanmar Surge in First Half of 2023," Reuters, July 20, 2023.

40 **It now supplies nearly half:** Kang, Milko, and Hinnant, "'The Sacrifice Zone.'"

40 **the militias have no legal rights:** "Myanmar's Poisoned Mountains," Global Witness.

40 **"Rare earths from Myanmar":** "Rare Earth Metals Used in Electric Vehicles May Come from Mines Controlled by Myanmar Junta," *Myanmar Now*, November 10, 2021.

41 **"In the areas we worked on":** "Quick Profits': Activists Fear for Environment under Military Rule," *Frontier Myanmar*, April 17, 2022.

CHAPTER 4: KILLING FOR COPPER

43 **this day in May of 2021:** I reconstructed this scene from details reported in local media, in particular: Tankiso Makhetha, "Coper Cable Thieves Get Violent and Deadly," *SowetanLIVE*, May 18, 2021; Tankiso Makhetha, "Families Heartbroken after Loved Ones Are Killed by Cable Thieves," *SowetanLIVE*, June 14, 2021.

44 **"We face these dangers":** Tankiso Makhetha, "Guard Recalls How His Two Colleagues Were Killed by 'Izinyoka,'" *SowetanLIVE*, June 14, 2021.

44 **will need to grow as much as threefold:** Saul Griffith, *Electrify: An Optimist's Playbook for Our Clean Energy Future* (Cambridge, MA: MIT Press, 2021), 2.

45 **No wonder Goldman Sachs:** Nicholas Snowdon, Daniel Sharp, and Jeffrey Currie, *Green Metals: Copper Is the New Oil* (New York: Goldman Sachs, April 13, 2021).

45 **Copper connects the cells:** Nick Pickens, Eleni Joannides, and Bhavya Laul, *Red Metal, Green Demand: Copper's Critical Role in Achieving Net Zero* (Edinburgh, UK: Wood Mackenzie, October 2022), 2–3.

45 **a typical EV contains as much:** "How Copper Drives Electric Vehicles," Copper Development Association, 2017, copper.org/publications/pub_list/pdf/A6192 _ElectricVehicles-Infographic.pdf.

45 **by 2035, global demand:** S&P Global, *The Future of Copper*, 9.

45 **By 2050, annual global demand:** S&P Global, *The Future of Copper*, 46.

45 **Analysts predict supplies will fall:** For example: Snowdon, *Green Metals*, 1–2.

45 **"Unless massive new supply":** S&P Global, *The Future of Copper*, 9–12, 25.

46 **copper mining has left:** Susan Schulman, "The $100bn Gold Mine and the West Papuans Who Say They Are Counting the Cost," *The Guardian*, November 1, 2016.

46 **In Mexico that same year:** Mike Holland, "Reducing the Health Risks of the Copper, Rare Earth and Cobalt Industries," *OECD Green Growth Papers*, no. 2020/03 (Paris: OECD, 2020), 48.

46 **"In Peru, police":** "Two Killed in Protest at $7.4-bln MMG Copper Project in Peru," Reuters, September 28, 2015, reuters.com/article/idUSL1N11Z05P/.

46 **In Pakistan, struggles:** S. Khan, "Why China's Investment Is Stoking Anger in Balochistan," *Deutsche Welle*, July 15, 2020, dw.com/en/why-chinese-investment -is-stoking-anger-in-pakistans-balochistan-province/a-54188705.

47 **In 1989, local residents:** "Update on the Panguna Mine," Rio Tinto, 2024, riotinto .com/en/news/trending-topics/panguna-mine.

47 **communities near the mine:** Ben Doherty, "Rio Tinto Accused of Violating Human Rights in Bougainville for Not Cleaning Up Panguna Mine," *The Guardian*, March 31, 2020; Ben Doherty, "After 32 Years, Rio Tinto to Fund Study of Environmental Damage Caused by Panguna Mine," *The Guardian*, July 20, 2021.

47 **Its English name comes via:** "Copper: An Ancient Metal," Dartmouth Toxic

Metals Superfund Research Program, sites.dartmouth.edu/toxmetal/more-metals /copper-an-ancient-metal.

47 **"The making of copper tools":** Mark Miodownik. *Stuff Matters: Exploring the Marvelous Materials That Shape Our Man-Made World* (Boston: Houghton Mifflin, 2014), 8.

48 **Archaeologists have unearthed:** Rob Tyson, "The Metals of Antiquity—Copper," Mining.com, March 30, 2021, mining.com/the-metals-of-antiquity-copper.

48 **The Bible gives a good sense:** Deut. 8:9 (New International Version).

48 **digging up copper in the Great Lakes:** "A Timeline of Copper Technologies," Copper Development Association, copper.org/education/history/timeline/timeline .html.

48 **The Haida people:** George F. MacDonald, *Haida Monumental Art* (Vancouver, BC: University of British Columbia Press, 1983), 6.

48 **Britain's Royal Navy:** S&P Global, *The Future of Copper*, 14.

49 **national telegraph systems:** "A Timeline of Copper Technologies."

49 **Telephones and electric power soon followed:** "A Timeline of Copper Technologies."

49 **French armorers came up:** "Copper's Millennia-Old Role in Conflict," Copper Development Association, Spring 2001, copper.org/publications/newsletters/discover /2001/Ct91/millennia.html.

49 **Global production tripled:** Lenka Muchová, Peter Eder, Alejandro Villanueva Krzyzaniak, *End-of-Waste Criteria for Copper and Copper Alloy Scrap: Technical Proposals* (Luxembourg: Publications Office of the European Union, 2011), 11.

49 **"A ruthless (perhaps even murderous)":** Timothy J. LeCain, *Mass Destruction: The Men and Giant Mines That Wired America and Scarred the Planet* (New Brunswick, NJ: Rutgers University Press, 2009), 17.

50 **killed hundreds of workers:** LeCain, *Mass Destruction*, 25; Kathleen McLaughlin, "Once-Powerful Montana Mining Town Warily Awaits Final Cleanup of Its Toxic Past," *The Washington Post*, February 10, 2020.

50 **Butte was the site:** LeCain, *Mass Destruction*, 12.

50 **"The environmental effects":** LeCain, *Mass Destruction*, 8.

50 **Ranchers near the Butte mines:** Diamond, *Collapse*, 3, 5.

50 **"Early miners behaved":** Diamond, *Collapse*, 5.

51 **only about 6 percent:** Daniel M. Flanagan, "Copper," *Metals and Minerals: US Geological Survey Minerals Yearbook 2018* (New York: US Government Publishing Office, 2018), 2.

51 **environmental groups and local Apaches:** Debra Utacia Krol, "Oak Flat: A Place of Prayer Faces Obliteration by a Copper Mine," *Arizona Republic*, August 18, 2021; Ernest Scheyder, "Rio Tinto's 26-Year Struggle to Develop a Massive Arizona Copper Mine," Reuters, April 19, 2021.

51 **Proposed mines in Alaska:** Jael Holzman, Ariel Wittenberg, and Hannah Northey, "Biden EPA Deals Major Blow to Pebble Mine," *E&E News*, May 25, 2022.

51 **"For decades, the bulk":** Javier Blas and Jack Farchy, *The World for Sale* (Oxford: Oxford University Press, 2021), 3–7.

51 **Mining supplies around 80 percent:** Pete Pattisson, "'Like Slave and Master': DRC Miners Toil for 30p an Hour to Fuel Electric Cars," *The Guardian*, November 8, 2021.

51 **Most Congolese subsist:** "The World Bank in DRC," The World Bank, September 25, 2023, worldbank.org/en/country/drc/overview.

51 **"On Sunday evenings":** Walter Isaacson, *Steve Jobs* (New York: Simon & Schuster, 2011), 38.

52 **"The farm became a hippy commune":** Sanderson, *Volt Rush*, 187.

52 **"Friedland went on TV":** Sanderson, *Volt Rush*, 189.

53 **That mine is a joint venture:** "Founder's Business Profile: Robert M. Friedland," Ivanhoe Capital Corporation, August 2019, ivanhoecapital.com/site/assets/files /4041/rmf-business-profile-august-2019.pdf.

53 **in exchange for far-ranging:** Jacob Kushner, *China's Congo Plan* (Washington, DC: Pulitzer Center on Crisis Reporting, 2013), 31.

53 **"We simply cannot continue":** Robert Friedland, Day 3 Keynote Presentation, CRU World Copper Conference.

53 **Chile contains the world's:** US Geological Survey, *Mineral Commodity Summaries 2022* (Reston, VA: US Geological Survey, January 2022), 54–55.

53 **where Indigenous people:** David R. Fuller, "The Production of Copper in 6th Century Chile's Chuquicamata Mine," *Journal of the Minerals, Metals & Materials Society* 56 (November 2004), 62–66, doi.org/10.1007/s11837-004-0256-6.

54 **Working conditions were grueling:** "Chuquicamata Mine," Encyclopedia.com, 2019, encyclopedia.com/humanities/encyclopedias-almanacs-transcripts-and-maps /chuquicamata-mine.

54 **Miners' wives waited:** "Copper: An Ancient Metal," *Dartmouth Toxic Metals.*

54 **"One would do well not":** Ernesto "Che" Guevara, *The Motorcycle Diaries* (New York: Verso, 1995).

55 **area the size of Manhattan:** Ed Conway, *Material World* (New York: Knopf, 2023).

55 **huge amounts of water:** Stephan Lutter and Stefan Giljum, "Copper Production in Chile Requires 500 Million Cubic Metres of Water," *Fineprint Brief No. 9* (Vienna: Vienna University of Economics and Business, December 2019); James Blair et al., *Exhausted: How We Can Stop Lithium Mining from Depleting Water Resources, Draining Wetlands, and Harming Communities in South America* (New York: Natural Resources Defense Council, 2022).

55 **Thieves have stolen copper slabs:** "Chilean Mining Players Urge More Forceful Measures against Copper Theft," *BNAmericas*, October 14, 2022.

55 **and wires from:** "How LatAm Telcos Grapple with Copper Cable Theft," *BNAmericas*, March 14, 2022; Juan Delgado, "Copper Thefts in Chile Linked to China," *Diálogo Américas*, March 19, 2020.

55 **Chile's train robbers:** Laura Millan, "Copper Cops Play Game of Cat and Mouse Around Desert Convoys," *Bloomberg*, February 15, 2019; James Attwood, "A Jump in Train Heists Has Chilean Copper Mines Turning to Trucks," *Bloomberg*, October 11, 2022; I also spoke with police officials in Chile who filled in some additional details. You can see one of these gangs at work in this video: T13, "Bandoleros del Norte: Robo de Cobre a Trenes y Fuertemente Armados," YouTube.com, January 9, 2023, youtube.com/watch?v=q2WjISq9Tb0.

56 **one of the most brazen heists:** Henry Shuldiner, "Chile's Copper Industry Under Siege as Robbers Attack Ports and Trains," *InSight Crime*, January 19, 2023.

56 **million-dollar nickel-thieving rings:** Dan Stamm, "Workers, Security Guards Swipe Nearly $1 Million in Nickel from Chester County Steel Plant: DA," NBC10

Philadelphia, December 14, 2016; US Department of Justice, "Baltimore Ware-house Owners Sentenced in Scheme to Steal $1 Million of Nickel Imported into the Port of Baltimore," press release, May 27, 2015.

56 **$10 million worth of cobalt:** Kit Chellel and Mark Burton, "Grand Theft Cobalt: Rotterdam," *Bloomberg Businessweek*, December 27, 2018.

56 **A particularly audacious operation:** Andy Hoffman and Bendikt Kimmel, "How Thieves Stole $40 Million of Copper by Spray-Painting Rocks in Turkey," *Bloomberg*, June 29, 2021.

56 **stolen in 2023 from Aurubis:** Julie Steinberg, "Massive Metals Theft Reported at Europe's Largest Copper Producer," *The Wall Street Journal*, September 1, 2023.

56 **In 2013, police in Arizona:** "ASARCO Multimillion-Dollar Copper Theft Ring Shut Down," *Arizona Daily Independent*, May 1, 2013. I also spoke with one of the investigators on the case.

57 **The booty includes fire hydrants:** Richard Gonzales, "Digging (Six Feet Under) for Scrap Metal," NPR's *All Things Considered*, October 6, 2008.

57 **a three-thousand-ton bell:** "Bell Heralds Break in Theft Case," Associated Press, March 7, 2008.

57 **a bust of Orville Wright:** Carrie Hodgin, "Wright Brothers Monument Theft: Copper Bust of Orville Wright Stolen from National Park," WFMY News 2, October 13, 2019.

57 **The US Department of Energy:** James A. Cook, "Beating the Red Gold Rush: Copper Theft and Homeland Security" (master's thesis, Naval Postgraduate School, December 2015), 21–22, apps.dtic.mil/sti/tr/pdf/ADA631979.pdf.

57 **at least one security guard:** "Copper Thief Sentenced to 15 Years in Thetford Mines Killing," CTV News Montreal, May 27, 2013.

58 **"Down below, temperatures can exceed":** Kimon de Greef, "The Dystopian Underworld of South Africa's Illegal Gold Mines," *The New Yorker*, February 20, 2023.

58 **Illegal miners have died:** Bheki Simelane and Greg Nicolson, "Blood and Gold: Zama Zamas Dice with Death in Daily Underground Hell," *Daily Maverick*, November 13, 2022.

58 **a mining company sealed off:** Poloko Tau, "Update: Eight Killed as Police Exchange Fire with Zama Zamas in North West," *City Press*, October 7, 2021.

59 **gangs have hijacked dozens:** Zandi Shabalala and Helen Reid, "Exclusive: Bandits Steal Truckloads of Copper Worth Millions in Southern Africa—Sources," Reuters, July 27, 2021.

59 **collaborate with actual power company:** Isaac Mangena, "City Power Is Turning the Tide on Copper Cable Thieves . . . ," City Power Johannesburg, press release, July 28, 2022, citypower.co.za/customers/Load%20Shedding%20Media%20updates/MEDIA%20STATEMENT%2028072022.pdf.

59 **more than $2 billion:** Myles Illidge, "This Crime Is Killing South Africa," *MyBroadband*, May 22, 2023.

59 **A Johannesburg hospital:** Sonri Naidoo, "Disastrous Theft Delays Reopening of Charlotte Maxeke," *The Star*, March 8, 2022.

59 **die of electrocution:** Noxolo Majavu, "City Power Records 24 Deaths Due to Electrocutions in Two Years," *City Press*, January 26, 2022.

60 **Police believe a rivalry:** Thabiso Malesele, "Five Murdered in Suspected Cable Theft Gang Rivalry in Eldorado Park," SABC News, January 19, 2022; Tankiso

Makhetha and Graeme Hosken, "Illegal Lesotho Miners' Rivalry at Centre of Tavern Bloodbath," *Sunday Times*, July 17, 2022.

60 **"This is the only language":** Maseo Nethanani, "Mob Justice Rules Over Copper and Cable Thieves in Muledane," *Polokwane Review*, March 9, 2023.

60 **four electric-company workers:** Koketso Ratsatsi, "Four Electricians Mistaken for Cable Thieves Killed," *SowetanLIVE*, March 9, 2023.

60 **The results are:** Gill Gifford, "Community Army Tackles Armed Cable Thieves," *The Herald*, April 21, 2022; Itumeleng Mafisa and Ntombi Nkosi, "Dudula Member Killed Protecting Eskom Workers after Cable Theft Incident," *The Star*, April 20, 2022.

CHAPTER 5: HOLDING POWER

63 **The price of nickel:** Marianne Lavelle, "Russia's War in Ukraine Reveals a Risk for the EV Future: Price Shocks in Precious Metals," *Inside Climate News*, March 28, 2022.

64 **"It is a very dangerous market":** Sam Meredith, "Nickel Prices Double to Record $100,000 a Ton, Trading Suspended in London," CNBC, March 8, 2022.

64 **An estimated five billion people:** Katrina Krämer, "The Lithium Pioneers," *Chemistry World*, October 17, 2019.

65 **three main parts:** Sanderson, *Volt Rush*, 26.

65 **corner of Mozambique:** Jael Holzman, "African Conflict Zone May Supply Key US Battery Material," *E&E News*, May 11, 2022.

65 **lithium ions flow from:** Scott Minos, "How Lithium-Ion Batteries Work," US Department of Energy, February 28, 2023, energy.gov/energysaver/articles/how-does-lithium-ion-battery-work.

65 **first began noodling:** "BU-204: How Do Lithium Batteries Work?," Battery University, February 22, 2022, batteryuniversity.com/article/bu-204-how-do-lithium-batteries-work.

65 **until the 1970s:** Krämer, "The Lithium Pioneers."

65 **a young Stanford researcher:** "M. Stanley Whittingham—Biographical," Nobel Prize.org, 2019, nobelprize.org/prizes/chemistry/2019/whittingham/biographical.

65 **"We came up with an idea":** Krämer, "The Lithium Pioneers."

66 **running on lead-acid batteries:** Sanderson, *Volt Rush*, 20, 27.

66 **less than 10 percent:** International Energy Agency, *Global EV Outlook 2022*, 8.

67 **"The battery plant just north":** Shannon Osaka, "The Unlikely Center of America's EV Battery Revolution," *The Washington Post*, April 17, 2023.

67 **Taken together, BloombergNEF estimates:** Adam Minter, "EV Battery Recycling Has Boomed Too Soon," *Bloomberg*, February 22, 2023.

67 **"As a latecomer":** Yergin, *The New Map*, 340–42.

67 **70 percent of all the world's:** International Energy Agency, *Global EV Outlook 2022*, 8.

67 **manufactures nearly one third:** Morgan Meaker, "The Rise and Precarious Reign of China's Battery King," *Wired*, June 28, 2022.

68 **more billionaires than Google or Facebook:** Sanderson, *Volt Rush*, 37.

68 **lavish government subsidies:** Sanderson, *Volt Rush*, 43–44.

68 **buy more electric vehicles:** International Energy Agency, *Global EV Outlook 2022*, 8.

68 **number-one drivers of:** International Energy Agency, *Critical Minerals Market Review 2023*, 7.

68 **The history of Norilsk:** Marianne Lavelle, "In the Russian Arctic, One of the Most Polluted Places on Earth," *Undark*, November 29, 2021; "Navigating the Transition to a Net Zero World," presentation, Nornickel Strategy Day, November 2021, 6, 7, 10–11.

68 **Humans have been using nickel:** "Nickel: Hidden in Plain Sight," Dartmouth Toxic Metals Superfund Research Program, sites.dartmouth.edu/toxmetal/more-metals/nickel-hidden-in-plain-sight.

69 **made mostly of copper:** "Nickel," US Mint Coin Classroom, usmint.gov/learn/kids/about-the-mint/nickel.

69 **The battery in a typical Tesla:** Lavelle, "Russia's War in Ukraine Reveals a Risk."

69 **The battery industry's consumption:** "Navigating the Transition to a Net Zero World," Nornickel Strategy Day, 46.

69 **a former Ministry of Foreign:** "Vladimir Potanin," Bloomberg Billionaires Index, bloomberg.com/billionaires/profiles/vladimir-o-potanin.

70 **"The Company produces metals":** "Momentum of Renewal," Nornickel Annual Report 2022, 3.

70 **"The company's pollution":** Lavelle, "In the Russian Arctic."

70 **Nornickel raked in $3.6 billion:** "Navigating the Transition to a Net Zero World," Nornickel Strategy Day, 11.

71 **were still flowing freely:** "World Nickel Statistics: Monthly Bulletin," International Nickel Study Group, October 10, 2023. The independent group Investigate Europe also reported, in October of 2023, that "Nornickel, the world leader in palladium and high-grade nickel, exported $7.6 billion worth of nickel and copper into the EU via Finnish and Swiss subsidiaries between the start of the war and July 2023": Pascal Hansens et al., "Russia: Europe Imports €13 Billion of 'Critical' Metals in Sanctions Blindspot," Investigate Europe, October 24, 2023, investigate-europe.eu/posts/russia-sanctions-europe-critical-raw-materials-imports.

71 **Russian nickel *increased*:** Eric Onstad, "Exclusive; EU, US Step Up Russian Aluminium, Nickel Imports since Ukraine War," Reuters, September 6, 2022.

71 **impending "large shortage":** The White House, *Building Resilient Supply Chains*.

71 **"Please mine more nickel":** Sanderson, *Volt Rush*, 155.

71 **written an open letter:** Aine Quinn, "Russian Arctic Peoples Appeal to Elon Musk for Nornickel Boycott," *Bloomberg*, August 7, 2020.

71 **Nirwana Selle was:** Rachel Cheung, "Workers Keep Dying at This Chinese Nickel Mining Company in Indonesia," *Vice*, February 7, 2023. I got more details from Selle's social media accounts on Instagram, which has since been closed, and TikTok (tiktok.com/@niisell55).

72 **two in a series of workers:** "Blaze at Indonesian Nickel Smelter Kills Worker, Wounds 6," Reuters, June 28, 2023.

72 **yet another fire:** Amy Chew, "Indonesia's Electric Battery Hub Bid Clouded by Mining Deaths," *Al Jazeera English*, July 11, 2023.

72 **left eighteen workers dead:** "Explosion at a Nickel Plant in Indonesia Leaves at Least 13 Dead and 46 Injured," *The Guardian*, December 24, 2023, theguardian.com/world/2023/dec/24/explosion-at-a-nickel-plant-in-indonesia-dead-injured;

"The Death Toll Rises to 18 in a Furnace Explosion at a Chinese-Owned Nickel Plant in Indonesia," Associated Press, December 26, 2023, apnews.com/article /indonesia-china-nickel-plant-explosion-e27a80240ab9376273ee906c3d8ecbdb.

72 **the government banned exports:** Sanderson, *Volt Rush*, 163.

72 **Investment has poured in:** Peter Yeung, "Workers Are Dying in the EV Industry's 'Tainted' City," *Wired*, February 20, 2023.

73 **they have collectively sunk:** Yudith Ho and Eko Listiyorini, "Chinese Companies Are Flocking to Indonesia for Its Nickel," *Bloomberg Businessweek*, December 15, 2022.

73 **Ford and other:** Jon Emont, "Ford Invests in $4.5 Billion Indonesia Facility to Secure Nickel for EV Batteries," *The Wall Street Journal*, March 30, 2023.

73 **leapt tenfold to $30 billion:** Peter S. Goodman, "How Geopolitics Is Complicating the Move to Clean Energy," *The New York Times*, August 18, 2023.

73 **twenty-one thousand acres of rainforest:** Yeung, "Workers Are Dying in the EV Industry's 'Tainted' City."

73 **Tailings and other mine waste:** Carolyn Fortuna, "Should Tesla Invest In Indonesia's Nickel Mines & Build a New Gigafactory There?," *CleanTechnica*, July 29, 2022.

73 **seeped into drinking water:** Febrina Firdaus and Tom Levitt, "'We Are Afraid': Erin Brockovich Pollutant Linked to Global Electric Car Boom," *The Guardian*, February 19, 2022.

73 **near mines in the Philippines:** Enrico Dela Cruz and Manolo Serapio Jr., "Philippines to Shut Half of Mines, Mostly Nickel, in Environmental Clampdown," Reuters, February 1, 2017.

73 **sparked violent clashes:** Nick Aspinwall, "Angry Philippine Islanders Are Trying to Stop the Great Nickel Rush," *Rest of World*, August 30, 2023.

73 **two basic forms:** Sanderson, *Volt Rush*, 162.

74 **runoff from acid-leaching refineries:** Rebecca Tan, Dera Menra Sijabat, and Joshua Irwandi, "To Meet EV Demand, Industry Turns to Technology Long Deemed Hazardous," *The Washington Post*, May 10, 2023.

74 **Chinese-owned nickel refinery:** Sanderson, *Volt Rush*, 155–58.

74 **spew out sulfur dioxide:** Antonia Timmerman, "The Dirty Road to Clean Energy: How China's Electric Vehicle Boom Is Ravaging the Environment," *Rest of World*, November 28, 2022.

74 **"complain about the dust":** Peter S. Goodman, "China's Nickel Plants in Indonesia Created Needed Jobs, and Pollution," *The New York Times*, August 18, 2023.

74 **Unrest over nickel:** Jonathan Barrett, "New Caledonia's Government Collapses over Independence, Nickel Unrest," Reuters, February 3, 2021.

74 **only a single nickel mine:** Garret Ellison, "Rising EV Demand Puts America's Only Nickel Mine in the Spotlight," MLive.com, July 16, 2022.

74 **stalled by lawsuits:** Walker Orenstein, "'Secrecy Is Unacceptable.' Minnesota Supreme Court Reverses NewRange Mining Permit after Regulators Shield Federal Criticisms," *MinnPost*, August 3, 2023.

75 **More than 70 percent:** US Geological Survey, *Mineral Commodity Summaries 2022*, 52–53.

75 **Democratic Republic of the Congo:** "The World Bank in DRC," The World Bank.

75 **"The titanic companies":** Siddharth Kara, *Cobalt Red* (New York: St. Martin's, 2023), 4–5.

75 **Men wearing nothing but shorts:** You can see this all for yourself in the many documentaries available on YouTube and elsewhere. For example: Unreported World, "Toxic Cost of Going Green," YouTube.com, October 31, 2021, youtube.com /watch?v=ipOeH7GW0M8.

75 **with hand tools:** Vivienne Walt and Sebastian Meyer, "The Race for Cobalt," Pulitzer Center, August 23, 2018, pulitzercenter.org/projects/race-cobalt.

76 **There, the ore is sorted:** Sanderson, *Volt Rush*, 106, 126, 128.

76 **"Children are required":** Benjamin K. Sovacool et al., "The Decarbonisation Divide: Contextualizing Landscapes of Low-Carbon Exploitation and Toxicity in Africa," *Global Environmental Change* 60 (January 2020).

76 **high levels of birth defects:** Sanderson, *Volt Rush*, 129–30, 137.

76 **exploitatively low pay:** RAID, *The Road to Ruin? Electric Vehicles and Workers' Rights Abuses at DR Congo's Industrial Cobalt Mines* (London: RAID, November 2021), 3, 4; Pattisson, "'Like Slave and Master.'"

76 **Amnesty International released:** Amnesty International, *Powering Change or Business as Usual?* (London: Amnesty International, 2023).

77 **Your smartphone probably contains:** Govind Bhutada, "The Key Minerals in an EV Battery," Elements.VisualCapitalist.com, May 2, 2022; Walt, "The Race for Cobalt."

77 **global production quadrupled:** The Faraday Institution, "Building a Responsible Cobalt Supply Chain," *Faraday Insights*, Issue 7 (May 2020, updated January 2023).

77 **the DRC was convulsed:** Andrew L. Gulley, "One Hundred Years of Cobalt Production in the Democratic Republic of the Congo," *Resources Policy* 79 (December 2022).

77 **an Israeli entrepreneur:** US Department of the Treasury, "Treasury Targets Corruption Linked to Dan Gertler in the Democratic Republic of Congo," press release, December 6, 2021.

78 **"made more money for himself":** Sanderson, *Volt Rush*, 105–106.

78 **Gertler has been investigated:** Willem Marx, "Forget Gas Prices. The Billionaire Club's Run on Cobalt Says Everything about Our Battery-Powered Future," *Vanity Fair*, April 21, 2022.

78 **founded in 1974:** Walker Orenstein, "How the Invasion of Ukraine Became Part of the Debate over Copper-Nickel Mining in Northern Minnesota," *MinnPost*, March 14, 2022.

78 **bribes in the DRC:** Sanderson, *Volt Rush*, 120.

78 **Managers in Beijing:** Stephen Chen, "Chinese Using a Mobile Phone in Beijing Effectively Manage Cobalt Mines in Africa by Remote Control: Study," *South China Morning Post*, December 7, 2022.

78 **artisanally mined cobalt:** Sanderson, *Volt Rush*, 149.

78 **more than half:** The Faraday Institution, "Building a Responsible Cobalt Supply Chain."

78 **sold on to companies:** RAID, "The Road to Ruin?," 6; Sanderson, *Volt Rush*, 136.

78 **2021 White House report:** Michael Holtz, "Idaho's Cobalt Rush Is Here," *The Atlantic*, January 24, 2022.

79 **"roughly a third":** Perzanowski, *The Right to Repair*, 33.

79 **US government has also accused:** Adam Morton, "Evidence Grows of Forced

Labour and Slavery in Production of Solar Panels, Wind Turbines," *The Guardian*, November 28, 2022.

79 **promised to reform:** Benjamin Hitchcock et al., *Recharge Responsibly: The Environmental and Social Footprint of Mining Cobalt, Lithium, and Nickel for Electric Vehicle Batteries* (Washington, DC: Earthworks, March 2021), 14–15.

79 **hiring third-party auditors:** "How the World Depends on Small Cobalt Miners," *The Economist*, July 5, 2022.

79 **Critics, however, charge:** Siddharth Kara, for one.

80 **"By then, the surrounding creeks":** Holtz, "Idaho's Cobalt Rush Is Here."

80 **at least one new cobalt mine:** As of March 2023, the mine was put on hold because of falling cobalt prices.

80 **given the DRC's titanic reserves:** US Geological Survey, *Mineral Commodity Summaries 2022*, 52–53.

80 **"There is no other work":** Kara, *Cobalt Red*.

81 **"The issue we're all addressing":** Krämer, "The Lithium Pioneers."

81 **Most of the world's phosphate rock:** International Energy Agency, *Critical Minerals Market Review 2023*, 41.

81 **"displace plants and animals":** "America's Frightening Phosphate Problem," Center for Biological Diversity, biologicaldiversity.org/campaigns/phosphate_mining/?ref=ambrook.

CHAPTER 6: THE ENDANGERED DESERT

84 **irreplaceable key ingredient:** International Energy Agency, *World Energy Investment 2022*, 127.

84 **all the lithium produced worldwide:** US Geological Survey, *Mineral Commodity Summaries 2022*, 100–101.

84 **the world will need:** International Energy Agency, *Critical Minerals Data Explorer*, updated July 11, 2023, iea.org/data-and-statistics/data-tools/critical-minerals-data-explorer.

84 **the world's biggest known lithium reserves:** US Geological Survey, *Mineral Commodity Summaries 2022*, 100–101.

84 **Brine operations in Argentina:** Amit Katwala, "The Spiralling Environmental Cost of Our Lithium Battery Addiction," *Wired*, August 5, 2018.

84 **Leaks from a Chinese:** Simon Denyer, "Tibetans in Anguish as Chinese Mines Pollute Their Sacred Grasslands," *The Washington Post*, December 26, 2016; Siyi Liu and Dominique Patton, "In China's Lithium Hub, Mining Boom Comes at a Cost," Reuters, June 14, 2023.

84 **Illegal miners have driven:** Linda Murjuru, "For Villagers in Zimbabwe, Lithium Boom Might Prove a Bust," *Global Press Journal*, September 5, 2023.

85 **a handful of niche products:** "The History of Lithium," International Battery Metals, December 14, 2021, ibatterymetals.com/insights/the-history-of-lithium.

85 **For the next thirty-odd years:** Taylor Quimby, "Outside/In: The Lithium Gold Rush," New Hampshire Pubic Radio, September 25, 2020, nhpr.org/environment/2020-09-25/outside-in-the-lithium-gold-rush#stream/0.

86 **The Atacama lithium mines:** Blair et al., *Exhausted*, 11.

86 **The Atacameño people:** Blair et al., *Exhausted*, 18.

86 **hundreds of gallons per second:** SQM, *Sustainability of Lithium Production in Chile* (Santiago, Chile: SQM, 2021), 9.

87 **one hundred thousand gallons:** Many estimates put this number much higher. See, for instance: Jorge S. Gutiérrez et al., "Climate Change and Lithium Mining Influence Flamingo Abundance in the Lithium Triangle," *Proceedings of the Royal Society B* 289, no. 1970 (March 9, 2022): 2; Blair et al., *Exhausted*, 10; Victoria Flexer, Celso Fernando Baspineiro, and Claudia Inés Galli, "Lithium Recovery from Brines: A Vital Raw Material for Green Energies with a Potential Environmental Impact in its Mining and Processing," *Science of the Total Environment* 639 (October 15, 2018).

87 **This process permanently shrinks:** "Yes, it's mining," said Corrado Tore, an SQM hydrologist who explained the process to me the day I visited the lithium mine. "We are reducing the amount of brine located in the reservoir."

87 *Engineering & Technology:* Ben Heubl, "Lithium Firms Depleting Vital Water Supplies in Chile, Analysis Suggests," *Engineering & Technology,* August 21, 2019.

87 **"has no material impact":** SQM, *Sustainability of Lithium Production in Chile,* 20.

87 **according to their analyses:** SQM, *Sustainability of Lithium Production in Chile,* 2, 22.

87 **each accused the other:** Dave Sherwood, "Water Fight Raises Questions Over Chile Lithium Mining," Reuters, October 18, 2018.

87 **A 2016 audit by Chile's:** Dave Sherwood, "Chile Indigenous Group Asks Regulators to Suspend Lithium Miner SQM's Permits," Reuters, September 13, 2021.

87 **forced it to cut back:** "Chile Lithium Producer SQM Gets Green Light on Environmental Plan," Reuters, August 30, 2022.

87 **The Chilean government filed:** Fabian Cambero, "Chilean State Sues BHP, Antofagasta Mines over Atacama Water Use," Reuters, April 8, 2022.

88 **used to be SQM's chairman:** Blake Schmidt and James Attwood, "Lithium King's $3.5 Billion Fortune Now Facing Government Threat," *Bloomberg,* June 23, 2022.

88 **make cash payments:** SQM, *Sustainability of Lithium Production in Chile,* 26–27.

90 **making the region hotter:** Gutiérrez, "Climate Change and Lithium Mining Influence Flamingo Abundance in the Lithium Triangle," 2; Brendan J. Moran et al., "Relic Groundwater and Prolonged Drought Confound Interpretations of Water Sustainability and Lithium Extraction in Arid Lands," *Earth's Future* 10, no. 7 (July 2022).

90 **Chilean government ordered BHP:** Cecilia Jamasmie, "BHP to Pay $93m for Environmental Harm at Escondida," Mining.com, June 4, 2021.

90 **study by Arizona State University:** Wenjuan Liu, Datu B. Agusdinata, and Soe W. Myint, "Spatiotemporal Patterns of Lithium Mining and Environmental Degradation in the Atacama Salt Flat, Chile," *International Journal of Applied Earth Observation and Geoinformation* 80 (August 2019), 1, 5, 6, 8, 10.

90 **study by a Chilean government agency:** Liu, Agustdinata, and Myint, "Spatiotemporal Patterns of Lithium Mining," 10.

90 **government study, published in 2018:** Sherwood, "Water Fight Raises Questions Over Chile Lithium Mining." Sherwood also posted the original document here: documentcloud.org/documents/5003677-Presentaci%C3%B3N-CORFO.html #document/p3/a461155.

92 **That would echo:** Heubl, "Lithium Firms Depleting Vital Water Supplies in

Chile": "E&T, in collaboration with satellite analytics firm SpaceKnow, has been able to produce further quantitative evidence that lithium brine mining efforts between 2015 and 2019 by SQM took a heavy environmental toll on a fragile water ecosystem within the Atacama salt flats. The analysis found a strong inverse relationship between water reservoir levels at SQM's ponds and the lagoons. As water levels in SQM's ponds increased, those in the lagoons would drop."

92 **study by Chilean and American researchers:** Gutiérrez et al., "Climate Change and Lithium Mining Influence Flamingo Abundance in the Lithium Triangle," 4, 6.

92 **the number of flamingos:** Gutiérrez et al., "Climate Change and Lithium Mining Influence Flamingo Abundance in the Lithium Triangle," 4, 5, 8.

93 **some eight thousand feet:** "CTR Commences Drill Program at Hell's Kitchen Lithium and Power," Controlled Thermal Resources, cthermal.com/latest-news /ctr-commences-drill-program-at-hells-kitchen-lithium-and-power.

93 **Colwell believes it could:** "The Power of California's Lithium Valley," *Controlled Thermal Resources*, cthermal.com/projects.

94 **including Berkshire Hathaway:** Janet Wilson and Erin Rode, "Lithium Valley: A Look at the Major Players Near the Salton Sea Seeking Billions in Funding," *Desert Sun*, May 13, 2022.

94 **The chronically impoverished:** "Unemployment Rate in Imperial County, CA," "Percent of Population Below the Poverty Level (5-Year Estimate) in Imperial County, CA," US Bureau of Labor Statistic, retrieved from FRED, Federal Reserve Bank of St. Louis, fred.stlouisfed.org/series/CAIMPE5URN, fred.stlouisfed.org /series/S1701ACS006025.

94 **Others have tried and failed:** For example: Kate Fehrenbacher, "Tesla Tried to Buy a Lithium Startup for $325 Million," *Fortune*, June 8, 2016.

95 **most raw lithium is sent:** Dolf Gielen and Martina Lyons, *Critical Materials for the Energy Transition: Lithium* (Abu Dhabi: International Renewable Energy Agency, January 2022), 6.

95 **have all signed on:** Max Schwerdtfeger, "BMW Joins Lithium Mining Project," *Mining Magazine*, February 28, 2022.

95 **audit by the Initiative for Responsible Mining Assurance:** "Mine Site Assessment Public Summary Report: SQM Salar de Atacama," Initiative for Responsible Mining Assurance, September 6, 2023, responsiblemining.net/wp-content/uploads /2023/09/Initial-Audit-Report-SQM-Salar-de-Atacama-FINAL-en.pdf.

96 **slash its water use:** SQM, *Sustainability of Lithium Production in Chile*, 22.

CHAPTER 7: DEPTH CHARGE

100 **known as polymetallic nodules:** "A Battery in a Rock. Polymetallic Nodules are the Cleanest Path Toward Electric Vehicles," The Metals Company, metals.co/nodules.

100 **twenty-one billion tons:** James R. Hein and Kira Mizell, "Deep-Ocean Polymetallic Nodules and Cobalt-Rich Ferromanganese Crusts in the Global Ocean," chap. 8 in *The United Nations Convention on the Law of the Sea, Part XI Regime and the International Seabed Authority: A Twenty-Five Year Journey* (Leiden: Brill, 2022), usgs.gov/publications/deep-ocean-polymetallic-nodules-and-cobalt-rich -ferromanganese-crusts-global-ocean-new.

100 **more of some metals:** Olive Heffernan, "Seabed Mining Is Coming—Bringing Mineral Riches and Fears of Epic Extinctions," *Nature*, July 24, 2019; James R.

Hein, Andrea Koschinsky, and Thomas Kuhn, "Deep-Ocean Polymetallic Nodules as a Resource for Critical Materials," *Nature Reviews Earth & Environment* 1 (February 24, 2020), 158–69.

101 **trigger an obscure legal process:** International Seabed Authority, "Nauru Requests the President of ISA Council to Complete the Adoption of Rules, Regulations and Procedures Necessary to Facilitate the Approval of Plans of Work for Exploitation in the Area," press release, June 29, 2021, isa.org.jm/news/nauru-requests -president-isa-council-complete-adoption-rules-regulations-and-procedures.

101 **one fifth of all the animal protein:** Jerry R. Schubel and Kimberly Thompson, "Farming the Sea: The Only Way to Meet Humanity's Future Food Needs," *Geo-Health* 3, no. 9 (September 2019), 238, 244.

101 **not to buy deep-sea metals:** United Nations Environment Programme Finance Initiative, *Harmful Marine Extractives: Understanding the Risks and Impacts of Financing Non-Renewable Extractive Industries* (Geneva: United Nations Environment Programme, 2022), 33–34.

102 *Hidden Gem's* **2022 trial:** The Metals Company, "NORI and Allseas Lift Over 3,000 Tonnes of Polymetallic Nodules to Surface from Planet's Largest Deposit of Battery Metals, as Leading Scientists and Marine Experts Continue Gathering Environmental Data," press release, November 14, 2022.

102 **The nodules have been growing:** "Polymetallic Nodules," International Seabed Authority, June 2022, isa.org.jm/files/documents/EN/Brochures/ENG7.pdf.

102 **One March day in 1873:** Kate Golembiewski, "H.M.S. Challenger: Humanity's First Real Glimpse of the Deep Oceans," *Discover*, April 19, 2019; John Murray, "The Cruise of the Challenger" first lecture, delivered in the Hulme Town Hall, Manchester, December 11, 1877.

102 **American geologist John L. Mero:** Michael Lodge, "The International Seabed Authority and Deep Seabed Mining," *Our Ocean, Our World* 54, nos. 1 & 2 (May 2017).

103 **James Bond–esque skullduggery:** "Project AZORIAN," Central Intelligence Agency, cia.gov/legacy/museum/exhibit/project-azorian.

103 **advancing marine technology:** Joshua Davis, "Race to the Bottom," *Wired*, March 1, 2007.

103 **a tennis buddy of Barron's:** Davis, "Race to the Bottom."

104 **half a billion dollars:** Justin Scheck, Eliot Brown, and Ben Foldy, "Environmental Investing Frenzy Stretches Meaning of 'Green,'" *The Wall Street Journal*, June 24, 2021.

104 **$30 million in profit:** Scheck, Brown, and Foldy, "Environmental Investing Frenzy Stretches Meaning of 'Green.'" Barron did not dispute this figure when we spoke.

104 **85 percent of the population:** Darian Naidoo, "Poverty and Equity Brief: Papua New Guinea," The World Bank, April 2020, databank.worldbank.org/data/download /poverty/33EF03BB-9722-4AE2-ABC7-AA2972D68AFE/Global_POVEQ_PNG .pdf.

104 **Barron came on as CEO:** Deep Sea Mining Campaign, London Mining Network, Mining Watch Canada, *Why the Rush? Seabed Mining in the Pacific Ocean*, July 2019, 13, deepseaminingoutofourdepth.org/wp-content/uploads/Why-the-Rush .pdf.

105 **granted permission to twenty-two companies:** "Exploration Contracts," International Seabed Authority, isa.org.jm/exploration-contracts.

105 **metals worth $31 billion:** Eric Lipton, "Secret Data, Tiny Islands and a Quest for Treasure on the Ocean Floor," *The New York Times*, August 29, 2022.

105 **A section of the treaty:** Agreement Relating to the Implementation of Part XI of the United Nations Convention on the Law of the Sea of 10 December 1982, multilateral, November 16, 1994, I-31364, 54–55, treaties.un.org/doc/Publication/UNTS/Volume%201836/volume-1836-I-31364-English.pdf.

106 **A single day's serious research:** WWF International, *In Too Deep: What We Know, and Don't Know, About Deep Seabed Mining* (Gland, Switzerland: World Wide Fund for Nature, 2021), 4.

106 **thirty-one marine researchers:** Diva J. Amon et al., "Assessment of Scientific Gaps Related to the Effective Environmental Management of Deep-Seabed Mining," *Marine Policy* 138 (April 2022).

106 **also interviewed twenty scientists:** Amon et al., "Assessment of Scientific Gaps Related to the Effective Environmental Management of Deep-Seabed Mining," 12.

107 **According to *Scientific American*:** Olive Heffernan, "Deep-Sea Mining Could Begin Soon, Regulated or Not," *Scientific American*, September 1, 2023.

107 **Corals, sponges, nematodes:** Philip P. E. Weaver and David Billett, "Environmental Impacts of Nodule, Crust and Sulphide Mining: An Overview," in *Environmental Issues of Deep-Sea Mining* (Cham, Switzerland: Springer Nature Switzerland, 2019), 23.

107 **Other creatures float:** Sabrina Imbler and Jonathan Corum, "Deep-Sea Riches: Mining a Remote Ecosystem," *The New York Times*, August 29, 2022.

107 **"I've been down there":** Todd Woody, "Explorer Victor Vescovo Says Deep Sea Mining Numbers Don't Add Up," *Bloomberg*, July 18, 2023.

107 **A recent study found:** Rob Williams et al., "Noise from Deep-Sea Mining May Span Vast Ocean Areas," *Science* 377, no. 6602 (July 7, 2022).

108 **A 2022 report:** UN Environment Programme Finance Initiative, *Harmful Marine Extractives*, 9–11, 22.

108 **eight hundred marine-science and -policy experts:** "Marine Expert Statement Calling for a Pause to Deep-Sea Mining," petition, seabedminingsciencestatement.org.

108 **"certain to disturb wildlife":** The Metals Company, Form S-1, US Securities and Exchange Commission, April 13, 2022, 20, 52.

109 **the first European ship:** Peter Dauvergne, "Dark History of the World's Smallest Island Nation," *The MIT Press Reader*, July 22, 2019.

109 **"moonscape of jagged limestone":** Anne Davies and Ben Doherty, "Corruption, Incompetence and a Musical: Nauru's Cursed History," *The Guardian*, September 3, 2018.

110 **moved all the detainees:** Ben Doherty and Eden Gillespie, "Last Refugee on Nauru Evacuated as Australian Government Says Offshore Processing Policy Remains," *The Guardian*, June 24, 2023.

110 **"Our people, land and resources":** Margo Deiye, "We Are the Forgotten Ones in the Climate Crisis, but Here's Our Solution," *The Independent*, December 10, 2022.

110 **"They were absolutely fucked":** The Metals Company, *Impact Report 2021* (Vancouver, BC: The Metals Company, 2022), 3, 107, metals.co/wp-content/uploads/2022/05/Final_MetalsCo_ImpactReport_052522.pdf.

111 **lost confidence in the enterprise:** Yusuf Khan, "Shipping Giant Maersk Drops Deep Sea Mining Investment," *The Wall Street Journal*, May 3, 2023.

111 **Journalists at *Bloomberg*:** Todd Woody at *Bloomberg* has reported extensively and deeply on this issue. For example: Todd Woody, "Mining Startup's Rush for Underwater Metals Comes with Deep Risks," *Bloomberg*, June 23, 2021.

111 **seemingly cozy ties:** For example: Louisa Casson et al., *Deep Trouble: The Murky World of the Deep Sea Mining Industry* (Amsterdam: Greenpeace International, 2020), 14–16.

111 **alleged that the ISA gave:** Lipton, "Secret Data, Tiny Islands and a Quest for Treasure on the Ocean Floor."

112 **three state-affiliated companies:** As of 2023, the three are: China Ocean Mineral Resources R&D Association (COMRA), China Minmetals Corporation, and Beijing Pioneer Hi-Tech Development Corporation.

112 **opposes the idea:** Jocelyn Trainer, "Geopolitics of Deep-Sea Mining and Green Technologies," United States Institute of Peace, November 3, 2022.

112 **Beijing contributes significant:** Lily Kuo, "China Is Set to Dominate the Deep Sea and its Wealth of Rare Metals," *The Washington Post*, October 19, 2023.

113 **a team of researchers:** "Metal-Rich Nodules Collected from Seabed During Important Technology Trial," DEME, April 22, 2021.

113 **"Something really bad":** Kris De Bruyne interview with the author, August 20, 2022.

115 **harvesting nodules by late 2025:** The Metals Company, "TMC Announces Corporate Update on Expected Timeline, Application Costs and Production Capacity Following Part II of the 28th Session of the International Seabed Authority," press release, August 1, 2023.

115 **The Norwegian government:** Gwladys Fouche and Nerijus Adomaitis, "Norway Moves to Open Its Waters to Deep-Sea Mining," Reuters, June 20, 2023.

115 **Japan has already:** Scott Foster, "Japan Dives into Rare Earth Mining Under the Sea," *Asia Times*, January 10, 2023.

115 **The Cook Islands:** Rachel Reeves, "When Deep-Sea Miners Come A-Courting," *Hakai Magazine*, July 25, 2023.

115 **Even Papua New Guinea:** John Cannon, "Deep-Sea Mining Project in PNG Resurfaces Despite Community Opposition," *Mongabay*, August 18, 2023.

CHAPTER 8: MINING THE CONCRETE JUNGLE

122 **Paul Revere, who melted:** Adam Minter, *Junkyard Planet* (London: Bloomsbury, 2013), 12.

122 **Today, scrap metal is:** "Scrap Metal Recycling in the US—Market Size, Industry Analysis, Trends and Forecasts (2024–2029)," IBISWorld, January 2024, ibisworld.com/united-states/market-research-reports/scrap-metal-recycling-industry/; Elizabeth S. Sangine, "Recycling—Metals," *Metals and Minerals: US Geological Survey Minerals Yearbook 2018* (New York: US Government Publishing Office, 2018), 1.

122 **Three quarters of all the lead:** Sangine, "Recycling—Metals," 3.

122 **one third of all the copper:** Pickens, Joannides, and Laul, *Red Metal, Green Demand*, 8.

122 **Millions of tons of metal:** "Ferrous Metals: Material-Specific Data," US Environmental Protection Agency, epa.gov/facts-and-figures-about-materials-waste-and-recycling/ferrous-metals-material-specific-data.

125 **"the largest lithium mine":** Charles Morris, "Tesla Cofounder JB Straubel: 'The Largest Lithium Mine Could Be in the Junk Drawers of America," *CleanTechnica*, May 2, 2021, cleantechnica.com/2021/05/02/tesla-cofounder-jb-straubel-the-largest -lithium-mine-could-be-in-the-junk-drawers-of-america/.

126 **millions of people are at work:** Kristin Hughes, "Waste Pickers Are Slipping Through the Cracks. Here's How We Can Support These Essential Workers During the COVID-19 Crisis," *World Economic Forum*, Sept 18, 2020.

126 **the people actually doing the work:** Rebecca Campbell et al., "From Trash to Treasure: Green Metals from Recycling," White & Case, May 5, 2022.

126 **In Pune, India:** "In India, Pune's Poorest Operate One of the World's Most Cost-Effective Waste Management Models," *WIEGO Blog*, February 20, 2019.

130 **Hundreds of American scrapyard workers:** Brian Joseph, "Recycling May Ease Your Conscience, but for Workers, It's Dirty, Dangerous and Even Deadly," *In These Times*, April 18, 2016.

130 **Two people died in 2014:** "Explosion at Scrap Metal Plant Kills Two," *Recycling Today*, August 28, 2014.

130 **a scrapyard worker in Arizona:** Phil Helsel, "Scrap Yard Worker Killed by Military Bomb Blast ID'd," NBC News, September 24, 2015.

130 **Eight children were:** Samy Magdy, "Blast Kills 8 Children Collecting Scrap Metal in Sudan," Associated Press, March 24, 2019.

130 **In Nigeria, Boko Haram:** Michael Olugbode, "Boko Haram Kill 55 Scrap Metal Collectors," *This Day*, June 12, 2022.

130 **In Afghanistan, decades of war:** Thomas Gibbons-Neff and Safiullah Padshah, "To Survive, Some Afghans Sift Through Deadly Remnants of Old Wars," *The New York Times*, May 14, 2022.

133 **Capital Salvage has even partnered:** It's called the Binners' Project, binnersproject .org.

133 **"Since the work was dirty":** Minter, *Junkyard Planet*, 31.

134 **ABC Recycling grew from the efforts:** For an intimate collection of memories about Jews in the Vancouver scrap industry, check out: Cynthia Ramsay, ed., *The Scribe: The Journal of the Jewish Museum and Archives of British Columbia* 34, Focus on the Scrap Metal Industry (Vancouver, BC: Jewish Historical Society of British Columbia, 2014).

136 **"The last of the American factories":** Minter, *Junkyard Planet*, 83.

136 **Minter visited one town:** Minter, *Junkyard Planet*, 7–8.

136 **"In China, lavish dinners":** Minter, *Junkyard Planet*, 58.

136 **One of the world's richest women:** David Barboza, "China's 'Queen of Trash' Finds Riches in Waste Paper," *The New York Times*, January 15, 2007.

137 **China's metal-recycling industry:** Berylin Cai, *Metal Recycling in China, Industry Report 4310*, IBISWorld, January 2022, 8, 29.

137 **"the Scrapyard to the World":** Minter, *Junkyard Planet*, 97.

137 **The intense heat involved:** "Violations at Metal Recycling Facilities Cause Excess Emissions in Nearby Communities," enforcement alert, US Environmental Protection Agency, July 2021.

137 **the carbon emitted per ton:** Moana Simas, Fabian Aponte, and Kirsten Wiebe, "The Future Is Circular: Circular Economy and Critical Minerals for the Green Transition" (Trondheim, Norway: SINTEF Industry, November 2022), 44–45.

CHAPTER 9: HIGH-TECH TRASH

140 **E-waste, as it's commonly known:** C. P. Baldé et al., *Global Transboundary E-Waste Flows Monitor 2022* (Bonn: United Nations Institute for Training and Research, 2022), 14.

140 **"a million 18-wheel trucks":** Perzanowski, *The Right to Repair*, 28.

140 **only 17 percent of all e-waste:** Baldé, *Global Transboundary E-Waste Flows Monitor 2022*, 15, 26.

141 **tens of thousands of e-waste scavengers:** Secretariat of the Basel Convention, *Where are WEEE in Africa? Findings from the Basel Convention E-Waste Africa Programme*, 2011, 6; Olakitan Ogungbuyi et al., *E-Waste Country Assessment Nigeria* (Basel, Switzerland: Secretariat of the Basel Convention, May 2012), 66.

142 **one registered mobile account:** "African Countries With the Highest Number of Mobile Phones," *FurtherAfrica*, July 19, 2022.

143 **5.3 *billion* mobile phones:** "International E-Waste Day: Of ~16 Billion Mobile Phones Possessed Worldwide, ~5.3 Billion Will Become Waste in 2022," *WEEE Forum*, October 13, 2022, weee-forum.org/ws_news/of-16-billion-mobile-phones-possessed-worldwide-5-3-billion-will-become-waste-in-2022.

146 **about five million tons:** Baldé, *Global Transboundary E-Waste Flows Monitor 2022*, 5, 8–9, 11.

146 **most of the e-waste in West Africa:** *Where Are WEEE in Africa? Findings from the Basel Convention E-Waste Africa Programme* (Basel, Switzerland: Secretariat of the Basel Convention, 2011), 6.

146 **In the United States and Europe:** *Strategy Paper for Circular Economy: Mobile Devices* (London: GSMA, 2022), 31.

147 **gear junked in Europe:** *Where Are WEEE in Africa?*, 18.

147 **75 percent of Nigeria's e-waste:** Ogungbuyi et al., *E-Waste Country Assessment Nigeria*, 3.

147 **In nearby Ghana:** *Where Are WEEE in Africa?*, 6.

149 **The smoke is thick and oily:** *Where Are WEEE in Africa?*, 6.

149 **informal e-waste scrapping:** A. A. Adeyi, B. Olayanju, and Y. Fatade, "Distribution and Potential Risk of Metals and Metalloids in Soil of Informal E-Waste Recycling Sites in Lagos, Nigeria," *Ife Journal of Science* 21, no. 3 (January 20, 2019).

150 **One ton of circuit boards:** Perzanowski, *The Right to Repair*, 37.

150 **"removed using highly corrosive acids":** Minter, *Junkyard Planet,* 184–86.

150 **Several studies of Guiyu:** Minter, *Junkyard Planet*, 185–86.

150 **study by Toxics Link:** Amita Bhaduri, "50,000 Workers Face Serious Health Risks in Illegal E-Waste Processing Units in Delhi," *Citizen Matters*, November 11, 2019.

151 **commissioned by Earthworks:** Hitchcock et al., *Recharge Responsibly*, 17.

151 **only about 5 percent:** Eric Frederickson, vice president of operations with Call-2Recycle, one of America's biggest battery recyclers, confirmed to me that this often-cited number is about as good an estimate as there is.

151 **batteries in everything:** Audrey Carleton, "Lithium Battery Fires Are Threatening Recycling as We Know It," *Vice*, February 1, 2022; Ryan Fogelman, "Li-Ion Battery Fires Unfairly Cost Waste, Recycling and Scrap Operators Over $1.2 Billion Annually," *Waste360*, February 3, 2021; "Industry, Consumer Steps Needed to Combat Scrap Yard Lithium-Ion Battery Fires," ScrapWare.com, June 26, 2022.

151 **in the United Kingdom:** Victoria Gill and Kate Stephens, "Batteries Linked to Hundreds of Waste Fires," BBC, November 30, 2022.

151 **Canada, and other countries:** Kevin Purdy, "Trashed Lithium-Ion Batteries Caused Three Garbage Truck Fires in California," *Ars Technica*, December 9, 2022.

151 **Fires sparked by faulty e-bikes:** I tallied these numbers from a startling number of local news reports.

154 **international regulations designed:** Adam Minter, *Secondhand* (London: Bloomsbury, 2019), 258.

155 **home to 80 percent:** International Energy Agency, *Critical Minerals Market Review 2023*, 47.

155 **120,000 tons of batteries:** "Battery Recycling and Utilization for the Development of a Circular Economy," CATL.com, December 31, 2022, catl.com/en/solution /recycling.

155 **investing billions in new plants:** Dan Gearino, "Inside Clean Energy: Here Come the Battery Recyclers," *Inside Climate News*, January 13, 2022.

155 **Umicore started as:** "History," Umicore.com, umicore.com/en/about/history.

155 **"It is clear that the biggest mine":** Heather Clancy, "Circular 'Mining' Reaches for the Mainstream," *GreenBiz*, March 7, 2022.

155 **nearly $2 billion in investment capital:** "Redwood Materials Raises Over $1 Billion in Series D Investment Round," RedwoodMaterials.com, redwoodmaterials .com/news/redwood-series-d.

156 **Canada-based Li-Cycle:** "Li-Cycle Is a Leading Global Lithium-Ion Battery Resource Recovery Company," Li-Cycle.com, li-cycle.com/about.

156 **Electric car batteries:** Casey Crownhart, "How Old Batteries Will Help Power Tomorrow's EVs," *MIT Technology Review*, January 17, 2023.

157 **one of the major difficulties:** Mark Burton and Thomas Biesheuvel, "The Next Big Battery Material Squeeze Is Old Batteries," *Bloomberg*, September 1, 2022.

158 **China has required manufacturers:** Blair et al., *Exhausted*, 27.

158 **According to CATL:** "Battery Recycling and Utilization for the Development of a Circular Economy," CATL.com.

158 **"there is currently no large-scale":** Simas, Aponte, and Wiebe, "The Future is Circular," 30, 31, 47.

158 **Less than 5 percent:** International Energy Agency, *World Energy Investment 2022*, 130.

159 **As for solar panels:** Anne Fischer, "There's Big Money in Recycling Materials from Solar Panels," *PV Magazine*, July 18, 2022; Jared Paben, "Solar Panels Are 'The New CRT' but Sector Is Preparing," *E-Scrap News*, May 13, 2021.

159 **most end-of-life panels:** Rachel Kisela, "California Went Big on Rooftop Solar. Now That's a Problem for Landfills," *Los Angeles Times*, July 15, 2022.

159 **China gives tax breaks:** Cai, *Metal Recycling in China*, 28.

159 **America spends billions:** Perzanowski, *The Right to Repair*, 236.

159 **Two major spending packages:** Justin Badlam et al., *The Inflation Reduction Act: Here's What's in It* (New York: McKinsey, October 2022), 8.

159 **as a work program:** "UNICOR Electronics Recycling," UNICOR.gov, unicor .gov/Recycling.aspx.

159 **American cars were once considered:** Minter, *Junkyard Planet*, 9, 162–67.

159 **in nearly all of its cars:** Minter, *Junkyard Planet*, 10.

160 **portable, inexpensive chemical reactors:** University of Leeds, "Mining Electronic Waste for Precious Metals," Medium, May 27, 2021.

160 **A Canadian startup:** Auxico Resources Canada, "Auxico Signs an MOU for the Exploitation and Trading of Rare Earths from Tin Tailings in Brazil," press release, January 6, 2022.

160 **The sap of *Pycnandra acuminata*:** Nina Notman, "The Magic Money Tree?," *Education in Chemistry*, April 25, 2022.

160 **Other plants can slurp up:** Ian Morse, "Down on the Farm That Harvests Metal from Plants," *The New York Times*, February 26, 2020.

161 **a company called Viridian Resources:** Ruth Longoria Kingsland, "Volunteers Battle Yellow-Tuft Alyssum in Southern Oregon," *Statesman Journal*, May 5, 2015.

161 **using recovered metals:** Elsa Dominish, Nick Florin, and Rachael Wakefield-Rann, *Reducing New Mining for Electric Vehicle Battery Metals: Responsible Sourcing Through Demand Reduction Strategies and Recycling*, report prepared for Earthworks by the Institute for Sustainable Futures, University of Technology Sydney, April 2021, 3, 6.

161 **less than 1 percent:** Dominish, Florin, and Wakefield-Rann, *Reducing New Mining for Electric Vehicle Battery Metals*, 4.

162 **"nothing—is 100 percent recyclable":** Minter, *Junkyard Planet*, 255.

CHAPTER 10: NEW LIVES FOR OLD THINGS

167 **The imperative to fix things:** Perzanowski, *The Right to Repair*, 17.

167 **"To reduce environmental impact":** Griffith, *Electrify*, 10.

167 **Extending the lifetime:** Federal Trade Commission, *Nixing the Fix: An FTC Report to Congress on Repair Restrictions* (Washington, DC: FTC, May 2021), 42–43.

168 **Ford and General Motors:** Elizabeth Evitts Dickinson, "Your Own Devices," *Harper's Magazine*, March 2022.

168 **Planned obsolescence, as it became known:** Perzanowski, *The Right to Repair*, 57.

168 **"In the nineteenth century":** Minter, *Secondhand*, 222.

169 **two hundred thousand Americans worked as:** Perzanowski, *The Right to Repair*, 26.

169 **The number of electronic:** "Electronic & Computer Repair Services in the US—Number of Businesses," *IBISWorld*, June 23, 2022, ibisworld.com/industry-statistics/number-of-businesses/electronic-computer-repair-services-united-states/.

169 **The electronics industry:** Federal Trade Commission, *Nixing the Fix*, 4, 7, 19–24.

169 **"The 2019 iMac manual":** Perzanowski, *The Right to Repair*, 108–109.

170 **Similar rules already cover:** Minter, *Secondhand*, 225, 234.

170 **one hundred right-to-repair bills:** Jack Monahan, "Our Picks for the Top Repair Stories of 2022," *Fight to Repair*, newsletter, January 15, 2023.

170 **On one side were:** Anne Marie Green, "Who Doesn't Want the Right to Repair? Companies Worth Over $10 Trillion," PIRG.org, May 3, 2021.

170 **"Apple told Nebraska lawmakers":** Perzanowski, *The Right to Repair*, 232.

171 **Apple CEO Tim Cook:** Tim Cook, "Letter from Tim Cook to Apple Investors," press release, January 2, 2019, apple.com/ca/newsroom/2019/01/letter-from-tim-cook-to-apple-investors.

171 **In 2021, the Federal Trade Commission:** Federal Trade Commission, *Nixing the Fix*, 7, 25, 27–32, 39, 55.

172 **By then, the European Union:** Federal Trade Commission, *Nixing the Fix*, 49–50.

172 **The UK also rolled out:** Cody Godwin, "Right to Repair Movement Gains Power in US and Europe," BBC, July 7, 2021.

172 **made a complete 180:** Dan Leif, "How Three OEMs Approach Product Sustainability," *E-Scrap News*, November 17, 2022.

172 **reporter tried out the service:** Brian X. Chen, "I Tried Apple's Self-Repair Program with My iPhone. Disaster Ensued," *The New York Times*, May 25, 2022.

172 **Lobbyists from Microsoft:** Kyle Wiggers, "New York's Right-to-Repair Bill Has Major Carve-Outs for Manufacturers," *TechCrunch*, January 3, 2023.

173 **In October of 2023:** Damon Beres, "Good News for Your Sad, Beaten-Up iPhone," *The Atlantic*, August 24, 2023.

173 **thousands of acres:** Kim Stringfellow, "Shifting Dust: Development and Demographics in Antelope Valley," MojaveProject.org, December 2017.

174 **used batteries are as much as:** Hauke Engel, Patrick Hertzke, and Giulia Siccardo, "SecondLife EV Batteries: The Newest Value Pool in Energy Storage," McKinsey.com, April 30, 2019.

174 **two million Nissan Leafs:** Office of Energy Efficiency & Renewable Energy, "How Much Power is 1 Gigawatt?," US Department of Energy, August 24, 2023, energy.gov/eere/articles/how-much-power-1-gigawatt.

175 **South Korean automaker Kia:** "Kia & DB to Reuse EV Batteries for Energy Storage," *Industry Europe*, September 7, 2022.

175 **Nissan is putting old:** Nissan et al., "Tennessee Partners Launch 'Second-Life' Battery Storage Project as Electric Vehicle Adoption Grows," press release, June 16, 2022, sevenstatespower.com/2022/06/20/tennessee-partners-launch-second-life-battery-storage-project-as-electric-vehicle-adoption-grows.

175 **Each year that they spend:** Melissa Ann Schmid, "Think before Trashing: The Second-Hand Solar Market Is Booming," *Solar Power World*, January 11, 2021.

176 **Human rights researchers charge:** Laura T. Murphy and Nyrola Elimä, *In Broad Daylight: Uyghur Forced Labour and Global Solar Supply Chains* (Sheffield, UK: Sheffield Hallam University Centre for International Justice, 2021).

176 **Inside a solar panel's cells:** Charlie Hoffs, "Mining Raw Materials for Solar Panels: Problems and Solutions," *The Equation*, October 19, 2022.

176 **contaminated rivers in Peru:** Christopher Pollon, *Pitfall: The Race to Mine the World's Most Vulnerable Places* (Vancouver, BC: Greystone, 2023), 82.

176 **In European countries:** Jared Paben, "Solar Panels Are 'The New CRT' but Sector Is Preparing," *E-Scrap News*, May 13, 2021.

176 **The International Renewable Energy Agency:** Paben, "Solar Panels Are 'The New CRT' but Sector Is Preparing."

176 **The US alone:** Taylor L. Curtis et al., *A Circular Economy for Solar Photovoltaic System Materials: Drivers, Barriers, Enablers, and US Policy Considerations* (Golden, CO: National Renewable Energy Laboratory Technical, 2021).

177 **the worldwide off-grid solar:** Lighting Global/ESMAP et al., *Off-Grid Solar Market Trends Report 2022: Outlook* (Washington, DC: World Bank, 2022).

177 **ten million used solar panels:** Adam Minter, "Used Solar Panels Are Powering the Developing World," *Bloomberg*, August 25, 2021.

177 **More than half of all the computers:** Minter, *Junkyard Planet*, 113.

177 **the used cell phone market:** "Secondary Mobile Market Tops $25b—Exceeding Demand or New Products—and Trend Set to Continue, Says B-Stock," RealWire,

January 30, 2019, realwire.com/releases/Secondary-mobile-market-tops-25b-and
-trend-set-to-continue-says-B-Stock.

177 **"Repair is skilled":** Perzanowski, *The Right to Repair,* 26.

178 **government subsidies could help grow:** Perzanowski, *The Right to Repair,* 236.

CHAPTER II: THE ROAD FORWARD AND HOW TO TRAVEL IT

179 **in October of 1971:** Peter Walker, *How Cycling Can Save the World* (New York: TarcherPerigee, 2017), 32.

180 **More than one out of every three:** Ralph Buehler and John Pucher, *Cycling for Sustainable Cities* (Cambridge, MA: MIT Press, 2021), 3.

180 **out of every two households:** Toon Zijlstra, Stefan Bakker, and Jan-Jelle Witte, *The Widespread Car Ownership in the Netherlands* (The Hague: Ministry of Infra-structure and Water Management, February 2022), 18.

181 **global clothing production:** Minter, *Secondhand,* 10.

181 **"Maybe fashion marketing":** J. B. MacKinnon, "The Price Is Wrong," *Sierra,* November 28, 2022.

181 **Manufacturing apparel generates:** MacKinnon, "The Price Is Wrong."

182 **a fifth of all food:** J. B. MacKinnon, *The Day the World Stops Shopping* (New York: Ecco, 2021), 4.

182 **America's 280 million–plus:** Mathilde Carlier, "Number of Motor Vehicles Registered in the United States from 1990 to 2022," Statista, February 28, 2024.

182 **single largest source:** Nadja Popovich and Denise Lu, "The Most Detailed Map of Auto Emissions in America," *The New York Times,* October 10, 2019.

182 **A 2013 study:** Edward Humes, "The Absurd Primacy of the Automobile in Amer-ican Life," *The Atlantic,* April 12, 2016.

183 **BloombergNEF estimates that:** Craig Trudell and River Davis, "EV Sales Will Triple by 2025 and Still Need More Oomph to Reach Net Zero," *Bloomberg,* June 1, 2022.

183 **conventionally powered vehicles:** "Oil Demand from Road Transport: Covid-19 and Beyond," BloombergNEF, June 11, 2020.

183 **research firm IHS Markit:** Yergin, *The New Map,* 414.

183 **account for at least half:** International Energy Agency, *Critical Minerals Market Review 2023,* 7.

183 **42,795 people died:** National Center for Statistics and Analysis, "Early Estimate of Motor Vehicle Traffic Fatalities in 2022," Report no. DOT HS 813 428, National Highway Traffic Safety Administration, April 2023, crashstats.nhtsa.dot.gov/Api /Public/ViewPublication/813428.

183 **they are pedestrians:** Governors Highway Safety Association, "New Projection: US Pedestrian Deaths Rise Yet Again in First Half of 2022," press release, February 28, 2023.

184 **Worldwide, the carnage is overwhelming:** Vince Beiser, *The World in a Grain* (New York: Riverhead, 2018), 68.

184 **more than 50 million people:** Etienne Krug, "Streets Are for People; It's Time We Give Them Back," World Health Organization, May 17, 2021, who.int/news -room/commentaries/detail/streets-are-for-people-it-s-time-we-give-them-back.

184 **more Americans than ever:** Environmental Protection Agency, "Explore the Au-tomotive Data Trends," epa.gov/automotive-trends/explore-automotive-trends-data.

184 **three hundred thousand tons of rubber:** Damian Carrington, "Car Tyres Produce Vastly More Particle Pollution than Exhausts, Tests Show," *The Guardian*, June 3, 2022.

184 **Low-income and nonwhite:** Thea Riofrancos et al., *Achieving Zero Emissions with More Mobility and Less Mining*, Climate and Community Project (Davis, CA: University of Califnornia, Davis, 2023), 9; Yoo Min Park and Mei-Po Kwan, "Understanding Racial Disparities in Exposure to Traffic-Related Air Pollution: Considering the Spatiotemporal Dynamics of Population Distribution," *International Journal of Environmental Research and Public Health* 17, no. 3 (February 2020).

184 **"the world's most underutilized asset":** Humes, "The Absurd Primacy of the Automobile in American Life."

185 **two billion parking spaces:** Jane Margolies, "Awash in Asphalt, Cities Rethink Their Parking Needs," *The New York Times*, March 7, 2023.

185 **two hundred square miles:** Henry Grabar, *Paved Paradise: How Parking Explains the World* (New York: Penguin Press, 2023), 12.

185 **twice as many parking spaces:** Benjamin Schneider, "The Bay Area Has Twice as Many Parking Spots as People—and There's a Hidden Toll," *San Francisco Examiner*, March 3, 2022.

185 **867 cars for every thousand people:** Yergin, *The New Map*, 414.

186 **"The exhilaration of bicycling":** Chris Carlsson, "19th Century Bicycling: Rubber Was the Dark Secret," *FoundSF*, foundsf.org/index.php?title=19th_Century _Bicycling:_Rubber_was_the_Dark_Secret.

186 **Marshall "Major" Taylor:** Michael Kranish. *The World's Fastest Man: The Extraordinary Life of Cyclist Major Taylor, America's First Black Sports Hero* (New York: Scribner, 2019).

186 **very first traffic accident:** Mary Bellis, "The Duryea Brothers of Automobile History," *ThoughtCo*, January 17, 2020, thoughtco.com/duryea-brothers-automobile -history-1991577.

187 **American bike sales plunged:** Hank Chapot, "The Great Bicycle Protest of 1896," *FoundSF*, foundsf.org/index.php?title=The_Great_Bicycle_Protest_of_1896.

187 **In a weird coincidence:** Owaahh, "'Nita Ride Boda Boda': How the Bicycle Shaped Kenya," *The Elephant*, March 28, 2019.

187 **regular cycling reduces:** Walker, *How Cycling Can Save the World*, 9.

188 **bike usage double:** Buehler and Pucher, *Cycling for Sustainable Cities*, 3, 10.

188 **In Taipei, Shanghai:** Buehler and Pucher, *Cycling for Sustainable Cities*, 5.

188 **Since Portland, Oregon:** Tim Henderson, "How Cities Learned to Love Bicycles," *Governing*, June 27, 2014.

188 **Bicycle commuting in LA:** Buehler and Pucher, *Cycling for Sustainable Cities*, 3, 16.

188 **one hundred million bikes:** "Bicycles Produced This Year," Worldometer, worldometers.info/bicycles.

188 **far outstripping car production:** "Cars Produced This Year," Worldometer, worldometers.info/cars.

188 **$70 billion global industry:** "Bicycles—Worldwide," Statista, statista.com/outlook /mmo/bicycles/worldwide#unit-sales.

188 **in Paris in 2007:** Buehler and Pucher, *Cycling for Sustainable Cities*, 8.

188 **into the millions today:** "The Meddin Bike-Sharing World Map," PBSC Urban Solutions, October 27, 2021, pbsc.com/blog/2021/10/the-meddin-bike-sharing

-world-map; Felix Richter, "The Global Rise of Bike-Sharing," Statista, April 10, 2018, statista.com/chart/13483/bike-sharing-programs.

188 **Hangzhou, China, has more:** Walker, *How Cycling Can Save the World*, 219.

189 **few large, state firms:** Amir Moghaddass Esfehani, "The Bicycle and the Chinese People," in *Cycle History: Proceedings of the 13th International Cycle History Conference*, eds. Andrew Ritchie and Nicholas Clayton (San Francisco: Rob van der Plas, 2003), 94–102.

189 **far outnumbering motor vehicles:** Neil Thomas, "The Rise, Fall, and Restoration of the Kingdom of Bicycles," MacroPolo.org, October 24, 2018.

189 **The government shifted support:** Thomas, "The Rise, Fall, and Restoration of the Kingdom of Bicycles."

189 **A massive shift in China's:** Walker, *How Cycling Can Save the World*, 149.

189 **Chinese bike-sharing industry's:** Thomas, "The Rise, Fall, and Restoration of the Kingdom of Bicycles."

190 **high-speed train network:** Ben Jones, "Past, Present and Future: The Evolution of China's Incredible High-Speed Rail Network," CNN, February 9, 2022.

190 **sales of e-bikes:** Christopher Mims, "The Other Electric Vehicle: E-Bikes Gain Ground for Americans Avoiding Gas Cars," *The Wall Street Journal*, April 6, 2022.

190 **France, for example:** Patricia Marx, "Hell on Two Wheels, Until the E-Bike's Battery Runs Out," *The New Yorker*, December 26, 2022.

190 **In the developing world:** David Wallace-Wells, "Electric Vehicles Keep Defying Almost Everyone's Predictions," *The New York Times*, January 11, 2023.

190 **More than two hundred million:** Mims, "The Other Electric Vehicle."

191 **worldwide market for electric vehicles:** Omar Isaac Asensio et al., "Impacts of Micromobility on Car Displacement with Evidence from a Natural Experiment and Geofencing Policy," *Nature Energy* 7 (October 27, 2022).

191 **To take an extreme example:** Riofrancos et al., *Achieving Zero Emissions with More Mobility and Less Mining*, 9; Park and Kwan, "Understanding Racial Disparities in Exposure to Traffic-Related Air Pollution."

191 **Half of all Americans:** Office of Policy Development and Research, "Urban. Suburban. Rural. How Do Households Describe Where They Live?," *PD&R Edge*, August 3, 2020, huduser.gov/portal/pdredge/pdr-edge-frm-asst-sec-080320.html.

191 **4.4 billion people:** "Urban Development," The World Bank, April 3, 2023, worldbank.org/en/topic/urbandevelopment/overview.

192 **"In the early 1900s":** Carlton Reid, "It's Been 100 Years Since Cars Drove Pedestrians Off The Roads," *Forbes*, November 8, 2022.

192 **"fully one third":** Clive Thompson, "The Invention of 'Jaywalking,'" *Marker*, March 28, 2022.

193 **Mehren's ideas, promoted:** Peter Norton, *Fighting Traffic: The Dawn of the Motor Age in the American City* (Cambridge, MA: MIT Press, 2008), 2.

194 **crashes in the Netherlands:** Henry Grabar, *Paved Paradise* (New York: Penguin Press, 2023), 294.

195 **more than triple:** "Cycling in the City," New York City Department of Transportation, nyc.gov/html/dot/html/bicyclists/cyclinginthecity.shtml.

195 **Paris has announced:** Grabar, *Paved Paradise*, 294.

195 **cut car use in half:** Saleem H. Ali, "There's No Free Lunch in Clean Energy," *Nature*, March 23, 2023.

196 **The Supreme court even proclaimed:** "Throughout the Rich World, the Young Are Falling out of Love with Cars," *The Economist*, February 16, 2023.

196 **driver's licenses is plummeting:** Andrew Van Dam, "The Oldest (and Youngest) States and the Shrinking Number of Teenagers with Licenses," *The Washington Post*, January 13, 2023.

196 **the rise of alternative means:** As cited in: "Throughout the Rich World, the Young are Falling out of Love with Cars," *The Economist*; Tim Henderson, "Why Many Teens Don't Want to Get a Driver's License," *PBS NewsHour*, March 6, 2017.

197 **"These are jobs that can't be":** Griffith, *Electrify*, 7–8.

197 **"Taxpayers subsidize motor vehicles":** Peter Dauvergne, *The Shadows of Consumption* (Cambridge, MA: MIT Press, 2008), 6.

197 **The International Monetary Fund:** David Coady et al., "Global Fossil Fuel Subsidies Remain Large: An Update Based on Country-Level Estimates," IMF Working Paper, no. 2019/089, May 2, 2019, 2.

198 **earmarks tens of billions:** Badlam et al., *The Inflation Reduction Act*, 3.

198 **it provides no:** Yonah Freemark, "What the Inflation Reduction Act Did, and Didn't Do, for Sustainable Transportation," Urban Institute, September 15, 2022.

199 **for transit e-buses:** Riofrancos et al., *Achieving Zero Emissions with More Mobility and Less Mining*, 9.

BIBLIOGRAPHY

This list below includes published books and a few key documents that I used in my research. Specific citations and other sources can be found in the Notes section.

Abraham, David S. *The Elements of Power: Gadgets, Guns, and the Struggle for a Sustainable Future in the Rare Metal Age*. New Haven, CT: Yale University Press, 2015.

Beiser, Vince. *The World in a Grain: The Story of Sand and How It Transformed Civilization*. New York: Riverhead, 2018.

Blas, Javier, and Jack Farchy. *The World for Sale: Money, Power, and the Traders Who Barter the Earth's Resources*. Oxford: Oxford University Press, 2021.

Bruntlett, Melissa, and Chris Bruntlett. *Curbing Traffic: The Human Case for Fewer Cars in Our Lives*. Washington, DC: Island, 2021.

Buehler, Ralph, and John Pucher. *Cycling for Sustainable Cities*. Cambridge, MA: MIT Press, 2021.

Conway, Ed. *Material World: The Six Raw Materials That Shape Modern Civilization*. New York: Knopf, 2023.

Crawford, Kate. *Atlas of AI: Power, Politics, and the Planetary Costs of Artificial Intelligence*. New Haven, CT: Yale University Press, 2022.

Dauvergne, Peter. *AI In the Wild: Sustainability in the Age of Artificial Intelligence*. Cambridge, MA: MIT Press, 2020.

———. *The Shadows of Consumption: Consequences for the Global Environment*. Cambridge, MA: MIT Press, 2008.

Diamond, Jared. *Collapse: How Societies Choose to Fail or Succeed*. New York: Viking, 2005.

Dorfman, Ariel. *Desert Memories: Journeys through the Chilean North*. Washington, DC: National Geographic, 2004.

Dunbar, W. Scott. *How Mining Works*. Englewood, CO: Society for Mining, Metallurgy, and Exploration, 2015.

Fisher, Jerry M. *The Pacesetter: The Untold Story of Carl G. Fisher*. Fort Bragg, CA: Lost Coast, 1998.

Grabar, Henry. *Paved Paradise: How Parking Explains the World.* New York: Penguin Press, 2023.

Griffith, Saul. *Electrify: An Optimist's Playbook for Our Clean Energy Future.* Cambridge, MA: MIT Press, 2021.

Hart, Matthew. *Gold: The Race for the World's Most Seductive Metal.* New York: Simon & Schuster, 2013.

Hund, Kirsten, Daniele La Porta, Thao P. Fabregas, Tim Laing, and John Drexhage. *Minerals for Climate Action: The Mineral Intensity of the Clean Energy Transition.* Washington, DC: World Bank, 2020. pubdocs.worldbank.org/en/961711588875536384/Minerals-for-Climate-Action-The-Mineral-Intensity-of-the-Clean-Energy-Transition.pdf.

Ilves, Erika, and Anna Stillwell. *The Human Project.* Self-published, Human Project, 2013.

International Energy Agency. *Critical Minerals Market Review 2023.* Paris: IEA, 2023. [License: CC BY 4.0] iea.org/reports/critical-minerals-market-review-2023.

Isaacson, Walter. *Steve Jobs.* New York: Simon & Schuster, 2011.

Kara, Siddharth. *Cobalt Red: How the Blood of the Congo Powers Our Lives.* New York: St. Martin's, 2023.

Khanna, Parag. *Connectography: Mapping the Future of Global Civilization.* New York: Random House, 2016.

Klinger, Julie Michelle. *Rare Earth Frontiers: From Terrestrial Subsoils to Lunar Landscapes.* Ithaca, NY: Cornell University Press, 2018.

Knowles, Daniel. *Carmageddon: How Cars Make Life Worse and What to Do About It.* New York: Abrams Press, 2023.

Kranish, Michael. *The World's Fastest Man: The Extraordinary Life of Cyclist Major Taylor, America's First Black Sports Hero.* New York: Scribner, 2019.

Kushner, Jacob. *China's Congo Plan: What the Economic Superpower Sees in the World's Poorest Nation.* Washington, DC: Pulitzer Center on Crisis Reporting, 2013.

LeCain, Timothy J. *Mass Destruction: The Men and Giant Mines That Wired America and Scarred the Planet.* New Brunswick, NJ: Rutgers University Press, 2009.

MacKinnon, J. B. *The Day the World Stops Shopping: How Ending Consumerism Saves the Environment and Ourselves.* New York: Ecco, 2021.

Marx, Paris. *Road to Nowhere: What Silicon Valley Gets Wrong about the Future of Transportation.* Brooklyn, NY: Verso, 2022.

McKibben, Bill. *Falter: Has the Human Game Begun to Play Itself Out?* New York: Henry Holt, 2019.

Minter, Adam. *Junkyard Planet: Travels in the Billion-Dollar Trash Trade.* London: Bloomsbury, 2013.

———. *Secondhand: Travels in the New Global Garage Sale.* London: Bloomsbury, 2019.

Murray, John. "The Cruise of the *Challenger.*" First lecture, delivered in the Hulme Town Hall, Manchester, December 11, 1877.

Norton, Peter D. *Fighting Traffic: The Dawn of the Motor Age in the American City.* Cambridge, MA: MIT Press, 2008.

———. "Street Rivals: Jaywalking and the Invention of the Motor Age Street." *Technology and Culture* 48, no. 2 (April 2007): 331–59. jstor.org/stable/40061474.

Penn, Robert. *It's All About the Bike: The Pursuit of Happiness on Two Wheels.* New York: Particular Books, 2010.

'erzanowski, Aaron. *The Right to Repair: Reclaiming the Things We Own.* Cambridge: Cambridge University Press, 2022.

Pitron, Guillaume. *The Dark Cloud: How the Digital World Is Costing the Earth.* Translated by Bianca Jacobsohn. London: Scribe, 2023.

———. *The Rare Metals War: The Dark Side of Clean Energy and Digital Technologies.* Translated by Bianca Jacobsohn. London: Scribe, 2020.

Pollon, Christopher. *Pitfall: The Race to Mine the World's Most Vulnerable Places.* Vancouver, BC: Greystone, 2023.

Ramsay, Cynthia, ed. *The Scribe: The Journal of the Jewish Museum and Archives of British Columbia.* Vol. 34, *Focus on the Scrap Metal Industry.* Vancouver, BC: Jewish Historical Society of British Columbia, 2014. jewishmuseum.ca/wp-content/uploads/2016/05/2014-SCRIBE_final_small-sz.pdf.

Sadik-Khan, Janette. *Street Fight: Handbook for an Urban Revolution.* New York: Viking, 2016.

Sanderson, Henry. *Volt Rush: The Winners and Losers in the Race to Go Green.* London: Oneworld, 2022.

Scheyder, Ernest. *The War Below: Lithium, Copper, and the Global Battle to Power Our Lives.* New York: Atria/One Signal, 2024.

Smil, Vaclav. *How the World Really Works: The Science behind How We Got Here and Where We're Going.* New York: Viking, 2022.

———. *Making the Modern World: Materials and Dematerialization.* Hoboken, NJ: Wiley, 2013.

Thwaites, Thomas. *The Toaster Project: Or a Heroic Attempt to Build a Simple Electric Appliance from Scratch.* New York: Princeton Architectural Press, 2011.

Veronese, Keith. *Rare: The High-Stakes Race to Satisfy Our Need for the Scarcest Metals on Earth.* Buffalo, NY: Prometheus, 2015.

Walker, Peter. *How Cycling Can Save the World.* New York: TarcherPerigee, 2017.

White House, The. *Building Resilient Supply Chains, Revitalizing American Manufacturing, and Fostering Broad-Based Growth.* 100-Day Reviews under Executive Order 14017. Washington, DC: The White House, June 2021.

World Bank Group. *Minerals for Climate Action: The Mineral Intensity of the Clean Energy Transition.* 2020.

Yergin, Daniel. *The New Map: Energy, Climate, and the Clash of Nations.* New York: Penguin Press, 2020.

INDEX